The BACKYARD GOAT

The BACKYARD GOAT

An Introductory Guide to Keeping
Productive Pet Goats

Sue Weaver

Storey Publishing

The mission of Storey Publishing is to serve our customers by
publishing practical information that encourages
personal independence in harmony with the environment.

Edited by Sarah Guare and Deborah Burns
Art direction and book design by Cynthia McFarland
Book layout by Karin Stack
Text production by Jennifer Jepson Smith

Cover photography by © Jason Houston (front), © John and Melody Anderson (back),
   and John Weaver (author)
Cover lettering by © Jim Datz
Part opener illustrations by © Jing Jing Tsong, Jing and Mike Company
Pen & ink illustrations by © Elayne Sears
Interior photography credits appear on page 206

Indexed by Christine R. Lindemer, Boston Road Communications

© 2011 by Sue Ann Weaver

Storey Publishing
210 MASS MoCA Way
North Adams, MA 01247
www.storey.com

Printed in the United States by Sheridan Books, Inc.
10 9 8 7 6

LIBRARY OF CONGRESS CATALOGING-IN-PUBLICATION DATA

Weaver, Sue.
 The backyard goat / by Sue Weaver.
    p. cm.
 Includes index.
 ISBN 978-1-60342-790-6 (pbk. : alk. paper)
 1. Goats. I. Title.
SF383.W33 2011
636.3'9—dc22
                          2010051170

Dedication

For MAC Goats Chief Forty-Five (Chiefee) —
you were a neat old goat.
And for Claudia Marcus-Gurn
and Emily Dixon,
who put good goats in my life.

Contents

Preface

Preface

My love affair with goats began later in my life, relatively speaking. My first love was horses. In fact, my mother says my first word was *horse*. Over the years I've shown, bred, and trained horses. I spent countless dollars buying and supporting the equine critters I dearly loved. I kept goats, because horses and goats go together like ice cream and pie, but goats weren't my life — until I met Chiefee.

Several years ago, I was asked to write about goats. As part of the research process, I subscribed to a passel of e-mail lists hosted by YahooGroups. Through them I met two women whose friendships have changed my life: Claudia Marcus-Gurn, of MAC Goats, and Emily Dixon, of Ozark Jewels.

That spring my husband, John, and I visited MAC Goats to collect a couple of three-week-old bottle-baby wethers the Gurns were graciously giving me and to photograph their beautiful Boers. Last stop on our tour of the Gurns' goat operation was the buck barn. When we ducked through the door, the biggest, brawniest old goat I'd ever seen ambled forward, propped his front feet on the bottom rail of his stall front, and fastened his soulful eyes on mine. In a heartbeat MAC Goats Chief Forty-Five, a.k.a. Chiefee, stole my heart. I knew I had to share my life with goats like him.

First came the bottle babies, Salem and Shiloh, whose mother was half sister to old Chiefee. Then a young buck from one of Chiefee's daughters joined our small herd, and before I knew it I owned eleven Boer goats! Soon I discovered packgoats and carting. And through Emily Dixon, I grew to love Nubians for their quirky personalities and yummy milk. Now twenty goats call our Ozark acreage their home, including my Boers, whom I treasure as my link to Chiefee.

And the horse obsession that burned so brightly for more than fifty years? It's still there, but considerably dimmed. I still love my horses, but goats (and sheep) are my passion now. I wouldn't have it any other way.

Part 1

Getting to Know Your Goats

CHAPTER 1

An Introduction to Goats

*If the beard meant everything,
the goat would preach.*

~ Danish proverb

The goat is among the oldest species domesticated by humans as a food source and for other purposes. Goats played a critical role in the lives of our distant ancestors, and, while their role today is not as widespread or integral to our survival, they still provide food, fiber, transportation, and companionship to countless people around the world. Their small size and grazing nature continue to make them the preferred livestock for the many farmers around the globe who must feed themselves on marginal land. To fully appreciate these special creatures, first learn about their history and how they function and think.

Origins of the Goat

For ancient races, goats were life-sustaining. Goats furnished food and carried packs. Bride prices and dowries were paid in goats, and goats provided sacrifices for the people's gods. In some cases, goats *were* the people's gods, or their gods had goaty attributes such as goats' legs and horns. Wealth was reckoned in goats — sometimes hundreds and thousands of them. The story of man and goat together is rich and diverse.

Ten thousand years ago, before any other livestock species had been tamed, villagers at Ganj Dareh, high in the Zagros Mountains of what is now Iran, made an alliance with wild goats. Humans would tame, feed, and protect the goats; the goats would, in turn, provide meat, milk, hides, hair, horns, bones, sinew, and dung for fuel.

Archaeological clues discovered at Ganj Dareh show proof of domestication. Goat bones are frequently unearthed at digs made in the Middle East, where hunters stalked wild goats since Neanderthal times. Typical bones recovered from food middens were mostly those of large old bucks. It made sense to target animals that packed the most meat. At Ganj Dareh, however, bones unearthed were mainly those of immature bucks. This, say goat domestication experts Dr. Melinda Zeder, curator of Old World Archaeology and Zooarchaeology at the National Museum of Natural History, and Dr. Brian Hesse, of the University of Alabama, was proof of domestication: does and a few mature bucks were kept for breeding, while extra bucklings and young adult males went into the family larder.

Meet the Bezoar Goat

The goats tamed at Ganj Dareh were Bezoar goats, also known as Bezoar ibexes, pasang, and *Capra aegagrus*. The Bezoar goat is a handsome and critically endangered wild goat still found in small numbers in the mountains of Asia Minor and the Middle East. It's also found on some Aegean Islands and on the island of Crete. These goats may represent relic populations of very early domestic animals that were taken to the Mediterranean islands during the prehistoric period and now live in feral populations.

Goats Like Company

Goats are happiest when they are in the company of another goat. A single goat usually copes if given another type of animal companion, preferably something close to its own size, such as a sheep or a miniature donkey. For the well-being of your goat, don't raise it alone.

Bezoar goats are slender mountain dwellers. They are light reddish brown in the summer and ash gray in the winter, with mahogany markings including eel stripes along their spines; flank and shoulder stripes; and dark markings on their legs, chests, tails, throats, and faces. Bucks are crowned by enormous scimitar-shaped horns that have sharp front edges and widely separated knobs running their length.

"Bezoar" also refers to a stonelike concretion formed by a swallowed mass of foreign material (usually hair or fiber) that collects in the stomach and fails to pass through the intestines. The ancients believed they formed only in the stomachs of Bezoar goats who were bitten by poisonous serpents and that they were a universal antidote against poisons. The word *bezoar* comes from the Persian word *pâdzahr,* which translates as "protection from poison."

Bezoar stones were highly prized and sold for fabulous sums as late as 1623, when, according to W. T. Fernie, writing in *Animal Simples Approved for Modern Uses of Cure* (see Resources), in a royal warrant sent to the Duke of Buckingham mention is made of "one great Bezar stone, sett in gould which was Queene Elizabethes."

Tracing the Ancestral Lines

Scientists soon discovered evidence of domestication at sites other than Ganj Dareh, and the early belief was that there were two ancestors of all goat breeds. In a 2001 study published in *The Proceedings of the National Academy of Science,* Gordon Luikart, of the Université Joseph Fourier in Grenoble, France, describes the findings of research into where domestica-

GOATS—WILD AND FERAL

While true wild goats still exist in our world (Bezoars, ibexes, and markhors, to name a few), what we call wild goats are usually feral goats whose ancestors were domestic at one time.

Among American feral breeds are Spanish and San Clemente Island goats, though few exist in the wild nowadays (the Rocky Mountain Goat, though a truly wild North American species, is a goat antelope, not a goat).

However, long-horned, shaggy feral goats, wild descendants of goats brought to England by Neolithic farmers, still exist in isolated pockets of mountainous and coastal areas of Scotland, Ireland, Wales, and England. Most are threatened with extinction, though conservators are working

A Bilberry British feral buck

hard to preserve them, particularly the stunning Bilberry and Old Irish goats of Ireland and the Lynton goats of Exmoor in England (see Resources, page 196).

---

tions occurred. The distant ancestors of most European and African breeds, Luikart and his colleagues found, were wild Bezoar goat does domesticated in the Fertile Crescent, but those of most Mongolian, Indian, and South Asian breeds were not. Their ancestor was a single doe tamed about nine thousand years ago in the Indus Valley of Pakistan.

In a later, unpublished study, geneticist Pierre Taberlet, one of Luikart's colleagues at Université Joseph Fourier, found that today's goats evolved through *five* distinct maternal lines, and that goats from the earliest two lines

arrived in southwestern Europe about seven thousand years ago, only three thousand years after the first domestication at Ganj Dareh.

By comparing the ancient goat mitochondrial DNA with modern goat DNA, they discovered that goats from both lines that originated in the Fertile Crescent were at the French site at the same time. Taberlet and his colleagues suspect that Neolithic farmers transported each line of goats into Europe along a separate westward route, one inland and the other running along the Mediterranean Sea.

PARTS OF A DOE

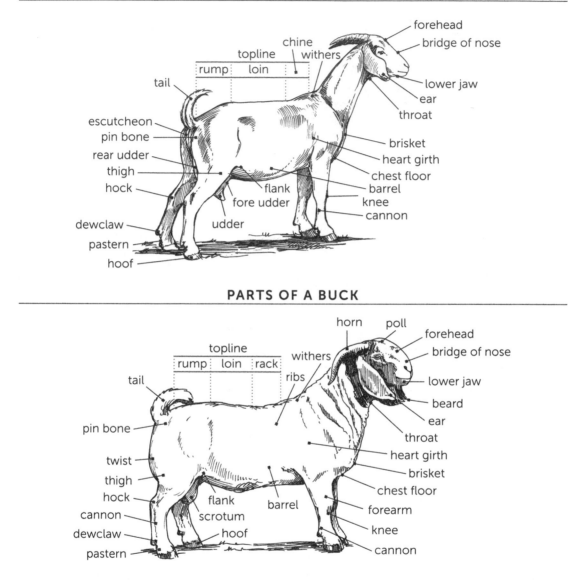

forehead
bridge of nose
chine
topline
withers
rump | loin
lower jaw
tail
ear
throat
escutcheon
pin bone
brisket
rear udder
heart girth
thigh
chest floor
hock
flank
barrel
fore udder
knee
dewclaw
cannon
udder
pastern
hoof

PARTS OF A BUCK

horn | poll
forehead
bridge of nose
topline
rump | loin | rack
withers
ribs
lower jaw
tail
beard
ear
pin bone
throat
twist
heart girth
thigh
brisket
hock
flank
chest floor
cannon
barrel
forearm
dewclaw
scrotum
knee
pastern
hoof
cannon

Goat Anatomy and Perception

Now that you have had a brief introduction to how goats evolved and why they were so important to our ancestors, perhaps you are starting to see why goats are so special. When you come face to face with a goat whose irrepressible personality melts your heart, like Chiefee did mine, you'll want to know everything about him. It also pays to understand his inner workings when you want to keep him happy and well. Here are some things to consider.

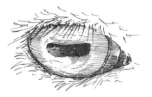

A goat's mysterious, slit-pupiled eyes give it a wider field of vision than ours.

Those Piercing Eyes

Have you ever looked into a friendly goat's eyes? Goats have horizontal, slit-shaped pupils that act like the panoramic lens on a camera, giving them an unusually wide field of vision. Other species such as cattle, deer, horses, and sheep have them too, but their dark brown irises cloak their rectangular pupils. Goats' pale irises (light brown, amber, or blue) don't.

Humans describe goats' eyes in terms ranging from wise and all-knowing to sinister and creepy. The Romans said their eyes were designed to see around corners. And goats stare. That alone discomfits many folks.

Sound the Horns

The bony appendages that grow from a goat's head are properly called horns, not antlers; horns are permanent structures while antlers are shed and grow back each year. (Though, as horns begin to grow on a young goat, you may see the outer surface peel; not to worry — this is normal and they will smooth out again.) Most goats, unless they're disbudded or dehorned, have horns. Some goats, however, are polled, meaning they never grow horns. According to a study conducted by Polish biologist Marcin Tadeusz Gorecki, horns are

the most important factor in determining a goat's place in his herd's pecking order. With goats, bigger *is* better.

Goat Physiology

LIFE SPAN: Usually 10 to 14 years (the known record is 23)

TEMPERATURE: 101.5 to 104.5°F (38.5–40.5°C)

PULSE: 70 to 90 beats per minute

RESPIRATION: 12 to 20 breaths per minute

RUMINAL MOVEMENTS: 1 to 3 per minute

RUMEN pH: 5.5 to 7.0

TEETH: Four pairs of lower incisors that line up with a dental pad on the upper jaw (goats have no upper front teeth), three premolars on each side of the upper and lower jaws, three molars on each side of the upper and lower jaws

STOMACH: A four-compartmented organ consisting of a rumen, a reticulum, an omasum, and an abomasum (see chapter 9)

INTESTINES: The small intestine is about 80 feet (24 m) long and has an average diameter of 1 inch (2.5 cm). It lies on the right side at the rear of the abdominal cavity.

CAECUM: The caecum marks the junction of the small and large intestines. It's about 8 inches (20 cm) long and 2 inches (5 cm) in diameter.

LARGE INTESTINE: The large intestine is about 16 feet (5 m) long.

Don't Mess with This Rack!

The largest set of goat horns listed in the *Guinness Book of World Records* belongs to Uncle Sam, a goat owned by William and Vivian Wentling of Rothsville, Pennsylvania. Uncle Sam's horns measured 52 inches (132 cm) from tip to tip.

The Bezoar goat, ancestor of domestic goats, has enormous, scimitar-shaped horns with a sharp front edge. A few types of goats still have them, but for the most part, horns became more oval and flattened under domestication. By 5000 BC some goats' horns were twisted, even corkscrew-shaped. Polled or dehorned goats appear in Egyptian art by 3000 BC.

Polled goats have large, round bumps on their head where horns would normally grow. A kid must have at least one polled parent to be polled. Although it seems to make sense to breed polled goats to other polled goats if you want to guarantee their kids won't have horns, this type of breeding is not without risk. Some kids born to two polled parents will be intersexed, meaning they'll have sexual traits that are neither fully male nor female. These goats are also called hermaphrodites.

Having horns can be beneficial to goats in hot weather. Horns act as thermoregulatory structures that help the goat wearing them to cool down. According to "Thermoregulatory Functions of the Horns of the Family Bovidae" (see Resources), "a typical goat at ambient temperature of 22°C [71.5°F] and in relatively still air can lose about 2 percent of its total

heat production through the horn." That's why packgoat and harness-goat owners usually let their goats' horns grow long. Long, stately horns make working wethers look regal too.

Nonetheless, horns do carry disadvantages. Horns are caught in woven-wire fences (nearly every horned goatling goes through this dangerous phase); horned goats hook and damage one another; horned heads don't fit in milking stanchions very well; and horned goats, especially bucks, are hard on their environment. A horned meat-breed buck can demolish a heavy-duty, welded-wire cattle panel overnight (night after night after night). Our Alpine wethers used their strong, scimitar horns like

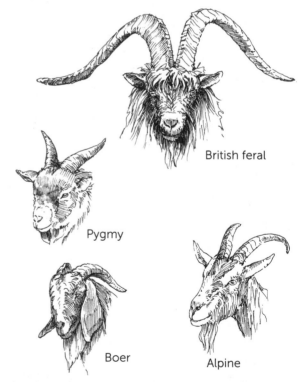

British feral

Pygmy

Boer

Alpine

Goat horns come in a wide variety of shapes and sizes, from the Boer and Pygmy doe's small spikes to the huge horns on Alpine and British feral goat bucks.

"Ziegen" ("Goats") heads are displayed in this fine old engraving from 1909. They are, clockwise from upper left, an ibex, a Bezoar goat, a domestic Angora goat, and a markor.

GOAT-SPEAK

The word *goat* comes from the Old English word *gat*, which meant "she-goat." In Old English a male goat was a *bucca*, which evolved into "buck," the correct term until a shift to *he-goat* (and *she-goat*) occurred in the late twelfth century.

"Nanny goat" originated in the eighteenth century and "billy goat" in the nineteenth. Nowadays the proper terms are *doe* and *buck*, and a castrated male is a *wether*.

claw hammers to destroy an entire fence line of expensive Red Top goat fence.

Because of these drawbacks, most kids are disbudded at a tender age: their horn buds are burned with a red-hot iron that destroys future horn growth. Dehorning adult goats is a cruel, gory process. Horns are living extensions of a goat's skull. Removing them causes massive bleeding and leaves gaping holes leading directly into a goat's sinuses; these take months to fill back in. If you don't want horned goats, don't buy them.

There are several things to remember when working with horned goats: Don't lead or restrain goats by their horns except in a dire emergency; they don't like it and broken horns bleed profusely. When you absolutely must restrain a goat by his horns, grasp both

This pretty French Alpine doe ("La Chèvre en Savoie") models a set of handsome wattles.

horns near their bases where they're strongest and don't exert more pressure than you have to. And remember, those horns can also hurt you. Don't lean over a horned goat, and when working closely with one, poke a small hole in each of two old tennis balls and force one onto the tip of each horn. When you're finished, pull them off and save them for another day.

Wattle It Be

Those globs of skin dangling on a goat's neck are wattles. Or, if you prefer, waddles, toggles, tassels, waggles, lassies, or cutaneous cervical appendices. Not all goats have them and they serve no known physiological purpose.

Pairs of wattles generally dangle from their wearer's throat, but they can crop up anywhere, particularly on the head or neck. They are found on both sexes and almost all breeds but are especially common on Swiss dairy goats, such as Alpines, Saanens, and Toggenburgs.

Wattles are thought to be inherited as a single dominant gene (bucks and does with wattles are likely to produce kids with wattles), and wattles can be a good thing. According to a study conducted in 1980, Saanen does with wattles produced 13 percent more milk than did those without.

Cute as they are, however, many breeders remove their goats' wattles. This is because kids sometimes suck on their peers' wattles, wattles are somewhat discouraged in the show ring, and they sometimes interfere with goats' collars. Another reason: wattle cysts occasionally develop at the base of a wattle. Wattle cysts may fill with clear fluid and are often mistaken for caseous lymphadenitis abscesses, but they are harmless.

Those who remove them usually wrap a snug rubber band around a newborn kid's wattles; with circulation restricted by the rubber band, the wattles fall off in a week or so. Other people simply snip them off. If you think you must remove your goat's wattles, restrain him, disinfect a pair of sharp scissors and the area at the base of each wattle; then, holding the wattle out away from your goat, snip it at the thin spot where it attaches to his body. Usually there isn't any bleeding.

A Goat's Five Senses

Ever wonder why you can never seem to be able to sneak up on a goat, or why some goats seem to be laughing (they're actually *flehmening*; see box on page 11)? Here's a look at how a goat perceives his environment.

Hearing. Goats have excellent hearing, particularly breeds with erect ears. Goats direct their ears in the direction of a sound (floppy-eared breeds lift the bases of their heavy ears to do this). They are sensitive to a wide range of sound — from the high-pitched shriek of a kid in peril to low-pitched snorts or hooves stamping on the ground.

Sight. Goats have prominent eyes, panoramic vision in the 320-to-340-degree range, and binocular vision of 20 to 60 degrees. Unless a goat is dozing or focused on something else, it's virtually impossible to sneak up on him or her. A German study conducted in 1980 showed that goats distinguish yellow, orange, blue, violet, and green from gray shades of equal brightness.

Smell. Goats have an acute sense of smell. Their olfactory system is more highly developed than our own. Scent helps bucks locate does in heat, and in turn draws does in heat to bucks. Scent (and taste) helps does recognize and bond with their newborn kids. They are more likely to move into the wind than with the wind, so they can better use their sense of smell.

Taste. Goats have more taste buds than we do (15,000 to our 9,000); they prefer certain tastes, particularly bitters, and are surprisingly selective about what they eat. Goats won't, for example, drink water fouled with feces or eat moldy or musty feed. They generally avoid poisonous plants unless the plants are wilted, which makes some species more tasty.

Touch. A goat's sense of touch is also acute. Flies alighting on goats elicit a strong response (especially in does on the milking stand).

He's Flehmening, Not Laughing

Your goat isn't laughing when he lifts his chin and curls his upper lip up and back; he's flehmening. By curling his lip, he exposes the vomeronasal organ (also called the Jacobson's organ), in the roof of his mouth, and draws scent toward it. This behavior helps him identify scents.

Bucks flehmen a lot, especially during rut, but does and wethers sometimes flehmen, too. Some goats flehmen more than others, though most goats flehmen when exposed to new, sharp scents. Thanks to the flehmen response, it's easy to teach goats to "laugh" on cue (in chapter 4 we'll show you how it's done).

Think Like a Goat

If you have a good sense of humor, patience, and a quick mind, you will love goats and they will love you. Still, it helps to understand why they do what they do. The best way is to pull up a bucket, sit down among your goats, and observe them on a regular basis, but these brief insights will help get you started right now.

Why Goats Fight

A defined pecking order exists in every herd, no matter its size. Where a goat stands in the hierarchy depends on age, sex, personality, aggressiveness toward other goats, and the size of (or lack of) his horns. Unweaned kids assume their dam's place in the order and usually rank immediately below her after weaning.

In a duel, goats rear and swoop down on their opponent.

Newcomers battle to establish a place in the herd. When fighting, males shove, butt, and side-rake opponents with their horns; does usually butt, jostle, and push.

Goats don't back up, lower their heads, and then race toward opponents in the manner of rams. Instead, they position themselves a few feet apart while facing one another, rear onto their hind legs, and then swoop forward, down, and to the side to smash their heads or horns against those of their foe.

Other forms of aggression toward other goats (or sometimes humans) are staring, horn threats (chin down, horns jutting forward), pressing horns or forehead against another goat, rearing without butting, and ramming an opponent's rear end or side. There is little infighting, however, once each herd member knows and accepts his place.

Herds are led by a herd queen, a tough old doe who has butted, shoved, and threatened her way to the top of the hierarchy. Other goats rarely (if ever) challenge her position. She remains herd queen until she leaves the herd or becomes too feeble to lead. When that happens, the position is often assumed by one of her daughters.

Most of the time males, including the herd king (the alpha buck), defer to the herd queen. During rut the herd king assumes leadership of the group. He breeds all the does; underling bucks aren't allowed to breed. Young, bold bucks, however, constantly challenge the herd king, so herd queens tend to outlast several kings.

Welsh Regimental Goats

BRITISH military mascots run the gamut from a drum horse named Winston to a ferret maintained by the 1st Battalion of the Prince of Wales's Own Regiment of Yorkshire. While they're nowadays kept for ceremonial purposes only, they once accompanied their units to the front.

Consider William Windsor, a.k.a. Billy, lance corporal in the 1st Battalion, the Royal Welsh, who recently retired with honors after eight years of mostly exemplary service. Like the other Welsh regimental goats, he had his own full-time handler, referred to as the Goat Major. His sole indiscretion occurred in 2006, when, in a parade honor-

ing Queen Elizabeth's eightieth birthday, he failed to keep in step and repeatedly tried to head-butt a drummer's nether parts, acts that resulted in his demotion to fusilier (the equivalent of private) for three whole months.

Three Welsh regiments kept white goat mascots until 2006, when they were combined into a single unit (the Royal Welsh) with William Windsor at its head. Since the mid-nineteenth century, all Welsh regimental goats have been Kashmir goats descending from broodstock presented to Queen Victoria (who loved pashmina cashmere scarves) by the Shah of Iran.

Shenkin (a.k.a. Taffy) was mascot of the 1st Battalion Royal Welch Fusiliers until it merged with two other Welsh units to become the Royal Welsh. Today a similar Shenkin leads the Regimental Band of the Royal Welsh on parade.

Here's a tip: The herd queen leads her herd to food, so if you lead your goats with a pail of feed instead of trying to drive them (which is a lesson in futility in any case), your goats will respect you as their queen. If you attempt to drive goats from the rear (bringing up the rear to protect the group is the herd king's job), they consider you their herd king — not a good position to be in when there are bucks in the herd.

Why Is My Baby Buckling Doing *That*?

When a buckling is born, his urethral process and the glans of his penis are attached to the inside of his prepuce (sheath) by a frenulum membrane. As his body begins producing testosterone — and that can happen when he's only a few weeks old — he begins "practice breeding" his dam and sisters. At the same time he may engage in a bizarre-looking form of air-humping behavior: his hind legs sink into a semi-squat and he repeatedly thrusts his hips. This helps break down the adhesion and enables him to extend his penis. When that happens, he's probably capable of impregnating females. This is why some goat breeders wean bucklings when they're only 8 weeks old, whereas doelings are generally left with their dams until they're at least 12 weeks of age.

What's on the Menu

Goats are browsers, not grazers. They prefer to range over a large area and eat a widely varied diet based on weeds, wild herbs, shoots, brush, twigs, and bark. Goats prefer to nibble the tops of plants instead of devouring them down to the ground. They won't eat plants contaminated with goat urine or droppings.

Feral goats are on the move, feeding, for up to 12 hours per day; domestic goats kept in large pastures or under range conditions feed this way too. When moving, goats tend to move in family units within a group. They spend their downtime sleeping or dozing and ruminating. Goats don't ruminate if they're nervous or on the alert.

Reading Body Language

When alarmed, goats curl their tails tight over their backs and snort a high-pitched, sneeze-like sound; they sometimes stomp a forefoot, too. Frightened goats flee a short distance and then turn to face whatever startled them. If pursued, they scatter.

When a goat is annoyed, the hair along his spine stands on end. His body hair sometimes stands up as well.

Goats pant when temperatures and humidity rise. Humidity, more than temperature, stresses goats.

Goats hate rain, water puddles, and mud. They readily move forward out of darkness and toward light. They move from confinement toward an open space, into the wind rather than downwind, and more readily uphill than down. They hate to cross water, dislike passing through narrow openings, and panic on slippery surfaces both natural and man-made (such as wet wooden milking stanchion decks and slick concrete). Goats have long memories and recall bad experiences for years.

When approaching wary goats, don't look directly into their faces, particularly their eyes. Predators do that. Instead, focus on the target goat's nearest shoulder.

The Buck Stops Here

Though regal and often affectionate to a fault, bucks have bizarre habits that make them unsuitable for most applications. Bucks, seasonal and aseasonal breeds alike, enter "rut" as autumn approaches. They stay in rut through the first months of winter. Bucks who live peaceably with other bucks the rest of the year become testy toward one another during rut; since many bucks (particularly bottle-raised bucks) consider humans part of their herd, they court female caretakers and challenge human males for leadership.

A two-hundred-pound buck is a force to reckon with, whether he's standing with his front feet on a woman's shoulders and blubbering in her face or ramming a man with his forehead or horns. People are seriously injured by bucks every year. If you're not an experienced goat keeper, don't keep a buck unless you really need one.

If the danger factor isn't enough, consider this: during rut, scent glands located near a buck's horns (or where his horns used to be) secrete incredibly strong-scented, greasy musk. When a buck rubs his forehead on a person or object, he's spreading his scent. Bucks become very vocal during rut; they're pretty hard to ignore. They also spray thin streams of urine along their bellies, on their front legs and chests, and into their mouths and beards. Bucks also twist themselves and grasp their penises in their mouths. They sometimes masturbate on their bellies and front legs and then sniff themselves and flehmen. Bucks don't make good pets or working goats.

· · · · · · · · · · · · ·

*Nobody gathers firewood
to roast a thin goat.*
~ Kenyan proverb

· · · · · · · · · · · · ·

2 Getting Your Goats

If you're short of trouble, take a goat.

~ Finnish proverb

You'll find that goats are fun, loving, intelligent, and marvelous friends. They also have an incredible array of uses. What type of goat do you want? Dairy does to milk? Pygmy goats for pets? A sturdy wether to pull a cart or tote a pack? Fiber goats whose fleece you can spin (or sell)? Are you looking for personality or production or both? How much money are you willing to spend? If you're not sure what you want, skip ahead to part 2, where I give detailed information about a goat's many roles. Write down what you want, think it over, and decide what you need before you buy.

Already know what you're looking for? Then you're probably wondering where to (and where not to) shop and how to evaluate the seller you're considering. And, of course, you're probably curious about what to look for, so I'll describe how to tell a healthy goat from a sick one.

Buying Your Goats

Let's assume you want a couple of registered Nubian dairy does. Where are you going to go to find them?

Check for GOATS FOR SALE notices on bulletin boards and in newspapers. You might see a notice pinned to a board at a feed store or veterinarian practice, or in the local classified ads, especially in penny saver–type shoppers. Or pin up a NUBIAN MILK GOAT WANTED sign of your own.

Talk to vets and county Extension agents in your buying area; they will likely know who is raising goats near your home. You may also want to take in a goat show. All state and most county fairs host them, and breed associations sanction them; e-mail or call organizations for dates and times. At shows, you can find information booths and talk to exhibitors after their classes.

Join some goat-related e-mail groups (see Resources). Breed-specific and general-interest goat lists host a "Friday sale" when subscribers post whatever they want to sell. It's a great way to find goats, as well as goat-related supplies, and you'll make new friends who just may own the goats you need. I met Emily Dixon that way, and all of my Nubians came from her farm!

Or you could visit breeders' websites. Type your breed, *goats,* and *sale* into your favorite search engine's search box (*Nubian goats sale*, for instance). If you like, qualify your search by state (*Nubian goats sale Missouri*). Alternatively, visit breed registry websites and consult their online member-breeder directories. If breeders' websites don't offer what you're searching for, e-mail and ask if they have it. If they don't, they may know someone who does.

Check out ads in goat magazines. Choose breed-specific journals or all-breed publications. Both types are packed with display ads, directories, and classifieds. Subscribe to your favorites or pick them up at a farm store. (See Resources for magazine and farm store information).

If the breed you need is a rare one, like San Clemente Island or Arapawa goats, log on to the American Livestock Breeds Conservancy website (see Resources) and click on *Classifieds* in the menu. Hard as it is to find these goats, they're there!

Where NOT to Buy Goats

The first rule of goat buying is to buy from individuals, not from a sale barn; sale barns are dumping grounds for sick goats and culls.

If you buy at the sale barn, you won't have access to a lot of important information about your goat: if she was vaccinated, if she's pregnant and by what sort of buck, or if there are any diseases or genetic conditions in her herd of origin. She might have any of several progressively degenerative, slow-incubating diseases such as caprine arthritic encephalitis (CAE; page 156) and Johne's disease (page 158). She or her herd mates might have foot rot or caseous lymphadenitis (CL) (page 156). Goats who weren't exposed to disease before they came to a sale barn will be by day's end.

If you attend such a sale, even just to look, you'll be tempted to buy. Regardless of whether you buy any goats, when returning home make sure to scrub your hands using plenty of soap

---

Buy a Sheep If You Want a Lawn Mower

Don't get goats to trim the grass in your yard. Goats are browsers, not grazers; they prefer leaves, twigs, and similar kinds of brush over grass. Goats also prefer to eat at shoulder height — perfect when browsing, not so great for mowing the lawn. For fun, lawn-mowing pets, buy sheep.

Goats will, however, clear brush with gusto. They'll happily strip your land of saplings and nuisance plants like poison ivy, multiflora rose, and blackberries. If brushing is your goal, you need goats!

and sanitize the clothes you wore to the sale before going near healthy goats or other livestock. Use one part household bleach to five parts plain water in a fine-mist spray bottle to spritz boots and shoes, and launder your other clothing in hot water and detergent. Foot rot, soremouth, respiratory diseases, and CL can hitchhike home on your hands and your clothes, so don't take chances.

Reality Check

Before you buy goats, make sure you really want them. They can be mischievous, stubborn, and often destructive. If you aren't positive you know what you're getting into, ask if you can spend a day with a friend's goats. Or volunteer at a goat dairy or book an overnight trip with a packgoat outfitter. Spend some time with goats before you commit. This is for the goats' sake as much as your own.

Make certain your way of life and goats will mesh. Are there zoning laws that might prevent you from keeping goats? Will neighbors complain? Are you able to provide housing, an exercise area or pasture, adequate fencing, feed, vet care, and hoof care for your goats? Goats are social creatures, so they crave the companionship of other animals. Plan on at least two goats or a goat and a sheep, llama, pet pig, or a horse or pony; you can't keep one goat all by himself.

And if you give in to temptation and buy, quarantine new goats from the sale barn for at least 30 days (see chapter 10 to learn how it's done).

How to Evaluate the Seller

Tap in to the local goat grapevine before you buy your goats. Ask goat owners which sellers they would buy from, which they would avoid, and why. Then, once you've narrowed the field to a handful of producers selling your type of goat, contact them and arrange to visit their farms.

The Visit

Be courteous and arrive on time. If you have goats or sheep at home and the seller wants to sanitize your shoes, don't be offended. Consider bio-security precautions a plus.

Look around. Goat farms don't tend to be showplaces, but they shouldn't be trash dumps, either. Are the goats housed in safe, reasonably clean facilities? Are the water tanks free of droppings and are goats eating hay in some sort of mangers up off the ground? Are the goats in good flesh, neither scrawny nor over-fat? In large herds, you'll spot a few goats who are skinnier or fatter than the norm, but the majority should be in average condition.

Ask about the seller's vaccination and deworming philosophies: Which vaccines and dewormers does he use and why? How often does she vaccinate and deworm her goats? Does he test for CAE, CL, or Johne's disease and does he have documentation? Are any of her goats currently infected? What about foot rot? Soremouth? If he's had these problems in his herd, what did he do to control them?

Does she show you just the goats you arranged to see or the entire herd? Try to see them all; if there are problems, you should know before you buy.

Ask why the goats are for sale. Is he changing bloodlines? Downsizing? Switching breeds? If they're culls, perhaps the trait he's culling for doesn't matter to you. For instance, maybe you don't need a high-producing dairy doe or don't mind an old Angora with coarser fleece. Or the goat whose strident bellowing drives her up a rope could be just the one you'd love to own. It happens!

The Papers

If you like what you see (we'll talk about evaluating goats later in this chapter), ask to examine the goats' registration papers and their health, vaccination, deworming, and production records.

Registered goats may be worth more money than ones who aren't, depending on the goats. If you're buying a doe and will sell her kids, papers make them much more valuable. With wethers it makes no difference, though; few goat associations register them at all.

When buying registered goats, carefully examine their papers to make sure you're getting what you pay for. Papers are "transferred" after every sale, so the papers should be issued in the seller's name. If they aren't, he in turn can't sign a transfer slip and the papers can't be transferred to you.

If you're buying a bred doe, you'll need a service memo signed by both the buck's and the doe's owners to register her kids. Depending on the registry you do business with, the service memo may be a separate document or part of the transfer slip.

Most registries stipulate that kids must be registered by their breeders. If you buy eligible but as yet unregistered kids, make sure to ask for a registration application that has been fully filled out and signed, as well as a completed transfer slip, which transfers their ownership to you.

Be aware that unethical or unknowing breeders sometimes sell goats that aren't quite what they seem. Purebred Boers, for example, are not the same as full-blood Boers; given two goats of equal quality, the full-blood is worth a lot more money. Learn the jargon before you go shopping. Each breed association has its own definition for breed terms; check out registration rules and terms at breed association websites.

Before handing over your check, ask for applicable guarantees and sales conditions in writing. Do this every time, even when dealing with friends.

Production records should indicate a goat's birth status (for example, single, twin, or triplet) and particulars about its reproductive career. Specifically ask about a doe's kidding habits: Has she had any birthing problems? Is she a good mom? Is the seller willing to work with you after the purchase should questions or problems arise?

Ask about guarantees. Some producers give them, some don't; if there is one and you buy, get it in writing.

Above all, trust your intuition. If the seller seems evasive or otherwise makes you feel uneasy, thank him for his time and look elsewhere. There are too many honest sellers in the goat world to deal with someone you don't quite trust.

Try Rescuing a Goat

There are a few more places to look for your goats. If your heart is big and you're looking for primarily pets instead of working animals, you could always rescue your goats. Any trip to a sale barn auction will turn up plenty of candidates. Doing this, however, is perilous for all the reasons mentioned in Where NOT to Buy Goats, not to mention that unhandled

Goat Tales

The Grateful Goat

ONCE UPON A TIME a Butcher bought a Goat; but as he was going to kill the Goat, and make him into meat for the table, the Goat opened his mouth and said —

"If you kill me, Butcher, you will be a few shillings richer; but if you spare my life, I will repay you for your kindness."

This Butcher had killed many Goats in his day, but he had never before heard them talk. Goats can talk to each other, as you must have heard; but most of them don't learn English. So the Butcher thought there must be something special about this Goat, and did not kill him.

The Goat felt very grateful that his life had been spared for a few more happy summers; and when he found himself free, the first thing he did was to go to the forest to see if he could find some means of repaying the Butcher's kind deed.

As he trotted under the trees, stopping now and then to crop some tender shoot that came within reach, he met a Jackal.

"I am glad to see you, Goatee," said the Jackal; "and now I'm going to eat you."

"Don't be such a fool," said the Goat. "Can't you see that I am nothing but skin and bones? Wait until I get fat. That's why I am here, just to get fat; and when I'm nice and fat, you may eat me and welcome."

The Goat was very skinny, in truth, and he pulled in his breath to make himself look more skinny. So the Jackal said—

"All right, look sharp, and be sure to look out for me on your way back."

"I shan't forget, Jackal," said the Goat. "Ta ta!"

By-and-by he fell in with a Wolf.

"Ha!" said the Wolf, smacking his lips; "here's what I want. Get ready, my Goat, for I am going to eat you."

"Oh surely not," said the Goat, "a skinny thing like me!" He drew in his breath again, and looked very skinny indeed. "I have come here to fatten myself, and when I am fat, you shall eat me if you like."

"Well," said the Wolf, "you don't look like a prize Goat, I grant you. Go along then, but look out for me when you come back."

"Oh, I shall look out for you!" said the Goat, and away he trotted.

By-and-by he came to a church. He went into the church, and there he saw last Sunday's collection plate full of gold coins. In that country anyone would be ashamed to put coppers into the plate, not because they or mistreated goats can be very, very hard to tame. That said, there is something so rewarding about saving a life. You could also check with any of the increasing number of privately run livestock rescue farms that are taking care of and rehabilitating animals that have been in unfortunate circumstances. Those farm owners are always on the lookout for good homes to take in their animals.

If you're after a pet or companion for your horse, or a harness or packgoat, consider raising a dairy-breed wether.

were rich, for they were not, but because they were generous. Now, Goats are not taught that they must not steal, but they think they have a right to whatever they can get hold of; so this Goat opened his mouth and licked up all the sovereigns, and hid them under his tongue.

The Goat went to a flower-shop, and asked the man who sold the flowers to make some wreaths, and to cover him up with them, horns and all. So the man covered him up with flowers, till he looked like a rose-bush. Then the Goat popped out a sovereign from his mouth, to pay the man, and very glad the man was to get so much money.

Then the Goat set out on his homeward way. He looked out for the Wolf, as he promised to do; and when the Wolf saw him coming along, he thought he was a rose-bush. The Wolf was not at all surprised to see a rose-bush walking along the road, for many were the strange things he had seen; and come to think of it, this was no stranger than a Goat that could speak English.

"Good afternoon, Rose-bush," said the Wolf; "have you seen a Goat passing this way?"

"Oh yes," said the Goat, "I saw him a few minutes ago back there along the road."

"Many thanks, Rose-bush," said the Wolf; "I am much obliged to you," and away he ran in the direction in which the Goat had come.

By-and-by he came to the Jackal.

"Hullo, Rose-bush!" said the Jackal. "Have you seen a Goat anywhere as you came along?"

"Oh yes," replied the Goat, out of the roses; "I saw him just now, and he was talking to a big Wolf."

"Good heavens!" said the Jackal. "I must look sharp if I want some Goat today," and off he galloped in a great hurry.

In the evening he got to the Butcher's house.

"Hullo!" said the Butcher, "what have we here?" He knew that rose-bushes could not walk, but he could not make out what it was.

"Baa! Baa!" said the Goat; "it's your grateful Goat, come back to pay you for your kindness." And with those words he spouted out all the sovereigns he found in the church, except for the one he paid the flower-man.

The Butcher was delighted to see so many sovereigns; he asked no questions, because he thought it wiser. He took the sovereigns, and found they were enough to keep him all his life, without killing any more goats. So he lived in peace, and the Goat spent his remaining years grazing comfortably in the Butcher's paddock.

— *The Talking Thrush and Other Tales from India,* William Henry Denham Rouse

Why Goats Climb on Cars

Goats climb on cars (and tractors and haystacks and the roof of your house or anything else that intrigues them) because they're there. The hundreds of breeds and types of goats in the world descend from mountain goats. Enough said?

The only reliable way to keep goats from climbing on something you'd rather they leave alone is to fence them out of the area. Barring that, arm waving and blasts from a high-pressure water gun like a Supershooter might help. But don't count on it.

An immutable fact about breeding goats is that half of all kids are males. Meat goat breeders don't mind; they sell full-blood and pure-bred bucklings as breeding stock and market the rest as slaughter kids. It isn't a huge problem for fiber goat folks, either; wethers produce outstanding fleece and sell well to hobby spinners. But dairy breed bucklings? There's a problem.

Dairy breed bucklings don't fatten quickly enough or well enough to raise and send to slaughter, and it's pound foolish to waste milk raising them for — what? So the sad fact is that compassionate breeders destroy surplus bucklings at or shortly after birth. Others send them to the sale barn as shoebox kids (newborns, so they're small enough to fit in a shoebox). A shoebox kid's future is grim. Most are purchased by ethnic buyers who use them as religious sacrifices, though a lucky few end up as pets. Even so, life is less than rosy; often deprived of colostrum and always exposed to sale-barn diseases, even these babies sometimes die.

Some breeders, however, give or sell these kids very inexpensively to responsible homes. Maybe you'd like one? If you think so, turn ahead to page 184 and read about raising a bottle kid. It's not a walk in the park, but it's ultra-rewarding and the ideal way to own a goat who loves you to the core.

To find a kid (or better, two, so that each has a companion from babyhood on), contact goat dairies and breeders in your locale, explain what you have in mind, assemble the necessary supplies, then sit back and wait till your kid is born. Keep in mind that responsible breeders would rather euthanize newborns than send them off to dubious lives as soon-to-be-discarded backyard pets, so be prepared to convince sellers that you are serious about providing a lasting home.

A girl smiles at an appealing kid in this vintage postcard image from France.

Teasel's Story

Even though it can be tough, rescuing a goat can be a rewarding experience. Let me tell you about our goat Teasel.

One of our neighbors raises slaughter kids from a herd of handsome mixed-Spanish does. One stood out among the rest: a cream-colored beauty with fantastic, spiral-twisted horns. One day we stopped to chat and, as usual, I admired that doe. Our neighbor said he was taking her to the sale barn the next week. She had had mastitis on several occasions and her udder was pendulous at best. In milk, her teats swung so close to the ground that her kids couldn't suckle and they died.

So, I handed him $60 (without looking closely at the rest of his stock, a mistake I came to rue) and purchased that doe. When we came back with our trailer, he lassoed her and pulled her to the trailer by her horns. Otherwise, she couldn't be touched.

At home she panicked if anyone came within 30 feet (9 m), so we quarantined her in a larger pen than usual and settled down to tame her. I named her Teasel because she was so pretty, yet oh so tough and wild.

The next thing we knew a lump came up on her throat. Oh no! So we put her in a box stall and continued working with her while the abscess came to a head. We lanced the lump, had it cultured, and yes, it was CL (caseous lymphadenitis).

We consulted a Texas veterinarian who specializes in autogenous vaccines. He advised us to vaccinate all of our goats, Teasel and the others alike, with an (expensive) autogenous vaccine, and to vaccinate the sheep with over-the-counter Case-Bac. Once Teasel's CL was under control, only she would require annual boosters. But before the vaccine arrived, another abscess developed and we went through the treatment protocol again. Surprisingly, Teasel became tamer in quarantine.

That was three years ago and Teasel is still wild (and abscess-free). She does, however, sneak up behind me when I'm feeding and poke me in the backside with her horns. She's a tough old bird and will never be tame, but we like her. Would we do it again? Yes, but one Teasel is enough.

Teasel

How to Choose a Healthy Goat

No matter what type of goats you need, buy healthy ones; don't make the mistake we made with Teasel. Consult the comparison chart (see page 26) for traits that should and shouldn't be present in the goats you'd like to buy. In particular, you'll want to consider the prospective goats' udder and teat conformation, as well as mouth structure.

Soft Udders and Two Teats

Dairy does should obviously have good udders, but meat and fiber breeds should, too. Kids can't nurse from the sort of badly deformed teats often seen in Boer goats or from udders that sag so low they nearly brush the ground.

Examine a doe's udder before you buy her. It should be globular, soft, pliable, and free of lumps. It should have good "attachments" (the ligaments that attach her udder to her body) that hold the udder up high and tight to her body. She should have two normal, shapely teats with no spurs or outgrowths — unless she's a Boer or Savanna; these breeds generally have four teats. Both sides of the udder should be the same size and both teats the same size or almost so.

When choosing a doeling who hasn't built her udder yet (it develops when she gives birth to her first kids), examine her mother's udder

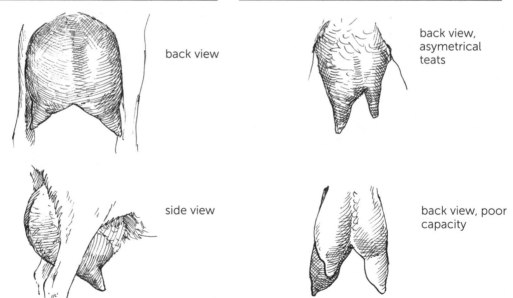

GOOD UDDERS	BAD UDDERS

back view

side view

back view, asymmetrical teats

back view, poor capacity

A good udder, like the one at the left, is balanced and capacious with shapely, symmetrical teats. It is held high and tight to the goat's body. Viewed from the side, one-third of an ideal, full udder bulges to the rear of the goat and another one-third is visible in front of the hind leg. The saggy udders at the right show asymmetrical teats and poor capacity; these are not productive udders.

sow mouth

perfect occlusion

parrot mouth

Select goats with a good mouth. Sow-mouthed and parrot-mouthed goats aren't able to browse or graze very well.

and her father's teats; this will give you a preview of what's to come.

Yes, her father's teats. Males have rudimentary teats in front of their scrotum (or where their scrotum used to be). It's important to check a buck's teats before buying his daughters or when choosing a mate for your doe. He should have the correct number of teats for his breed, and they shouldn't be misshapen.

A Normal Mouth

A goat's lower teeth should be flush with her upper dental pad when her mouth is closed. If her lower teeth extend beyond the dental pad, she is "sow-mouthed" or "monkey-mouthed"; this is fairly common in Roman-nosed (arch-faced) breeds such as Boers and Nubians, and it's severely penalized in the show ring.

If her dental pad extends beyond her lower teeth, she is "parrot-mouthed." This is severely

penalized as well. And worse, badly parrot-mouthed or sow-mouthed goats have problems browsing and grazing. They tend to lose weight on pasture and need hay and possibly grain to survive.

Both sow-mouth and parrot-mouth conditions are hereditary and should be avoided, especially in young goats, as both of them worsen with age.

Other Factors to Consider

Most goats are healthy, but you'll want to make sure your goats have been tested for the major diseases: caprine arthritic encephalitis (CAE), caseous lymphadenitis (CL), and Johne's disease. I'll discuss these in chapter 10. Also, you'll want to choose individual goats based on your particular needs, taking the following factors into consideration.

HOW TO IDENTIFY SOUND GOATS

HEALTHY GOATS	SICK OR INJURED GOATS
• Are alert and curious. • Stand in a normal pose with their head up and their tail carried straight back or slightly up (in the case of some dairy breeds) or flipped forward over their back	• Are dull and uninterested in their surroundings • Frequently isolate themselves from the rest of the herd • Often stand hunched over with their tail hanging down • Grind their teeth
• Have mucous membranes that are bright pink and moist	• Have pale mucous membranes, which are associated with anemia and a heavy internal parasite load; blue, yellow, or red membranes indicate disease
• Have bright, clear eyes	• Have dull, depressed-looking eyes • Have an opaque film over one or both eyes, an indication of blindness or pinkeye • Have a clear discharge (tearing) from one or both eyes • Have fresh or crusty opaque discharge globbed in the corners of their eyes
• Have a dry, cool nose (a trace of clear nasal discharge isn't cause for concern) • Breathe regularly and without labor *Note:* Goats issue a sneeze alarm when startled; this is normal	• Have thick, opaque, creamy white, yellow, or greenish nasal discharge • Wheeze, cough, sneeze, or breathe heavily and/or erratically
• Have teeth that are normal for their age • Have normal-smelling breath (remember: goats burp gas) • Have an injury- and abscess-free mouth and tongue	• Young goats have broken or missing teeth • Have putrid-smelling breath (possibly caused by abscessed teeth) • Have sores in their mouth
• Have blemish-free lips	• Have fluid-filled pimples or blisters, which usually indicate soremouth — it's very contagious, even to humans
• Have nonirritated, properly shaped ears (refer to breed standards)	• Shake their head, indicating an irritant in one or both ears, possibly parasites or weed seeds (foxtail awns are notoriously hard to remove)
• Have a clean, glossy hair coat and pliable, vermin- and eruption-free skin	• Have a dull, dry hair coat; skin may show evidence of external parasites or disease

HEALTHY GOATS	SICK OR INJURED GOATS
• Move freely and easily	• Move slowly, unevenly, or with a limp • Hold up or refuse to put weight on a leg
• Have normal-size joints	• May have swollen joints (typical of goats with CAE)
• Weigh the average amount for their breed and age	• May be thin or emaciated (goats with CAE and Johne's disease become increasingly emaciated as these fatal diseases progress, so never buy sick-looking, skinny goats)
• Have a healthy appetite (most goats love food and will mob you for it)	• Won't eat
• Ruminate (chew cud) after eating	• A goat who doesn't cud is very ill
• Have firm, berrylike droppings; tail and surroundings areas are clean	• Has scours (a.k.a. diarrhea); tail, tail area, and hair on hind legs are matted with fresh or dried diarrhea
• Produce clear urine	• Produce cloudy or bloody urine
• Does have a soft, even, blemish-free udder with normal teats and produce normal-looking milk that tests negative using the home California Mastitis Test (CMT) (see page 183)	• Have a hard, lumpy, or discolored udder with injured or abnormally swollen teats • Have a rash or pustules, which may indicate soremouth or rain rot • Have blood-streaked milk or milk containing clots of blood; watery or clumped milk
• Wethers and bucks have a normal sheath enclosing a normal penis	• Have blood or crystals on the hair around the opening of the sheath, a swollen sheath, and dribbling urine
• Have a normal temperature (101.5–104.5°F [38.5–40.5°C] for adult goats; slightly higher values are acceptable for kids)	• Run a high or low temperature; a subnormal temperature is generally more worrisome than is a fever

A healthy goat is bright-eyed and alert. A sick goat is hunched over, with its tail down.

healthy

sick

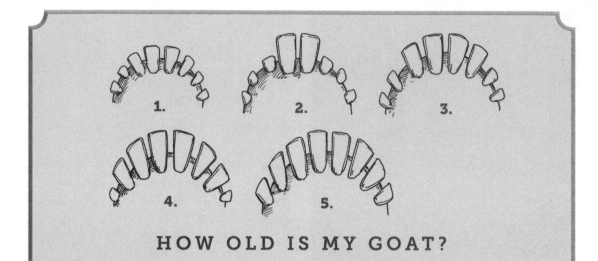

HOW OLD IS MY GOAT?

It's easy to estimate a goat's age by examining the eight teeth in the front of its mouth.

1. **Less than 1 year:** All the teeth are sharp and small (they are baby teeth).
2. **Between 1 and 2 years:** The center two teeth are big (these are permanent teeth) and the rest are small.
3. **Roughly 2 years:** The four center teeth are big and the rest are small.
4. **About 2½ years:** The six center teeth are big and the rest are small.
5. **About 3 years:** All eight teeth are big.
Older than 3 years: The teeth gradually spread and may eventually fall out in old age.

Age

If you're looking for pets, working wethers, or a companion for your horse or pony, think about starting with kids. Kids bond with the humans who raise them and the type of animals they find in their environment. However, take extra care with kids. Build separate quarters in your horse's stall until the little guys are big enough not to be stepped on, and give packgoat and harness kids time to grow up before putting them to work.

Harness goats shouldn't be driven and packgoats shouldn't carry weight until they're at least 2 years old. If you're eager to pack or drive right now, choose a more mature goat.

Dairy doelings are fun to raise but not so much fun to train to the milk stand. Also, first-fresheners (does who have recently given birth to their first kids) give less milk than older does and have teeny, hard-to-milk teats. If you haven't milked before, you're much better off with a seasoned adult.

In Angora-type goats (Angoras, Pygoras, and Nigoras), the younger the better, as fiber coarsens with age. You don't want old goats if you need fine fiber for spinning, though their fleece is excellent for making dolls' wigs and Santa beards.

How old is old? It depends. Most does and wethers live 12 to 14 years; some live much longer. Does should, however, be retired from kidding when they're about 10 years old, as kidding problems increase with age.

Sex

I gave a warning about bucks in chapter 1, but it bears repeating: If you don't breed goats, you probably won't want a buck.

If you milk, obviously you need does. And while harness goats and packgoats are traditionally wethers, if you have a doe and don't breed her, she can be a working goat, too. Some packers take low-production dairy does in milk out on the trail to provide fresh milk for meals. If you plan to do this, choose a doe with a small, well-attached udder, not one with an easily injured, floppy udder and dangly teats.

Wethers make the best all-around goats because they aren't preoccupied with kids, heat cycles, or rut. Unless you plan to milk or breed, think wethers.

Suitability

Choose goats that can do the job you have in mind. For instance, Myotonic goats (unless they rarely faint) and Nigerian Dwarves don't make good backcountry packgoats. But be objective when choosing. If you don't need a lot of milk, non–dairy breed does may provide all you need; some Spanish and Boer goats produce almost as much cashmere as bona fide Cashmere goats; and if you day-pack on fairly level terrain, a Pygmy goat can carry your sandwiches.

To learn more about evaluating goats, ask the organization that registers your breed of choice for a copy of its breed standard. This is a list of points to look for when judging that breed of goat. Have fun goat shopping!

. . . .

*Do not mistake
a goat's beard for a
fine stallion's tail.*
~ Irish proverb

. . . .

Bill the Navy Goat

IN 1893, midshipmen of the United States Naval Academy, in Annapolis, Maryland, put their heads together and decided their football team needed a mascot. A goat, they thought, would be just the thing. With this in mind, they borrowed El Cid, mascot of the USS *New York*, and took him to the Army-Navy game.

Navy won the game, so when the USS *New York* was in port for the 1900 match, they borrowed El Cid again. They painted his horns blue and gold and decked him out in a fine blue blanket emblazoned with **NAVY** in golden lettering on both sides. After Navy's 11-7 victory, El Cid accompanied the midshipmen back to Annapolis. Somewhere along the line they renamed him Bill.

Since then, a long line of Bills have led the Navy team. Most of them have been Angora goats. The most famous of all was also known as Three-to-Nothing Jack Dalton, who is stuffed and on view in the Navy field house. He served from 1906 through 1912 and got his name because twice during his tour of duty, Navy's Jack Dalton kicked field goals for 3-0 victories over Army.

Today's Bill is the 34th to bear the honorable name. A group of midshipmen known as Team Bill volunteer their time to care for Bill and his understudies and to transport Bill to nearby games. Navy also has a costumed human mascot to spare the real live Bills the stress of traveling long distances.

Since 1890, Bill the goat has served as mascot to the midshipmen of the United States Naval Academy. All latter-day Bills have been Angora goats like this handsome fellow (left), but early Bills were also other breeds (right).

The Breed You Need

*Look for the black goat
while it's still daylight.*

~ Nigerian proverb

No matter what trait you are seeking in a goat, chances are you can find the breed that fulfills your needs. Want to have fresh goat's milk? A goat to pack with on long trails? A friendly pet that will enjoy being with your children? There are more than 30 goat breeds in the United States alone and scores of breeds in lands across the sea. Which breed tickles your fancy?

The following goats are all available in North America. Americans have access to a diverse range of breeds, from the ubiquitous Boer to the critically endangered Arapawa and San Clemente Island goats. A few, such as Kinders, Myotonics, and Miniature Silky goats, were pioneered by American breeders. The rest are immigrants, though most of the breeds in the following section have been here long enough to be as American as the Fourth of July.

ALPINE AND MINI-ALPINE

Also called the French Alpine.
Type: Dairy
Origin: The French Alps (Alpines are, however, considered a Swiss breed.)
Color: See Alpine Colors, page 32.

A Mini-Alpine goat

Description: Alpines are productive milkers, hardy, agile, uncommonly friendly, and easily adaptable to a variety of climates. The breed originated in Switzerland; however, in 1922, Dr. Charles P. DeLangle imported from France the first documented herd of Alpine goats (18 does and three bucks) to reach the United States, hence their secondary name: French Alpines.

Colorful Alpine goats are medium-large animals with erect ears, a straight or dished face, and a short to medium-length coat. Mature females measure at least 30 inches (76 cm) at the withers and should weigh at least 135 pounds (61 kg). Males should measure from 34 to 40 inches (86–102 cm) and weigh not less than 170 pounds (77 kg).

Adult Mini-Alpine does must be at least 23 inches (58 cm) tall and no more than 29 inches (74 cm). Mini-Alpine bucks must be, at minimum, 24 inches (61 cm), and maximum, 31 inches (79 cm).

Alpine Colors

Alpine goats, both full-size and miniature, come in an array of attractive colors with interesting French names.

cou blanc (*coo blahnc*) or **"white neck."** White front quarters and black hindquarters with black or gray markings on the head

cou clair (*coo clair*) or **"clear neck."** Front quarters are tan, yellow, off-white, or shading to gray with black hindquarters

cou noir (*coo nwah*) or **"black neck."** Black front quarters with white hindquarters

sundgau (*sund-gow*). Black with white markings, such as underbody or facial stripes

pied. Spotted or mottled

chamoisee (*sham-wah-zay*). Brown or bay; characteristic markings are black face, feet, legs, dorsal stripe, and sometimes a martingale running over the withers and down to the chest. Spelling for male is *chamoise*.

two-tone chamoisee. White front quarters with brown or gray hindquarters; this is not a cou blanc or cou clair, as these terms are reserved for animals with black hindquarters

broken chamoisee (sundegau, cou blanc, as examples). Solid chamoisee broken with another color by being banded or splashed with white

ANGORA

TYPE: Fiber
ORIGIN: Turkey
COLOR: Angoras are white; colored Angoras are black (ranging in shade from coal black to silver), faded red, and shades of brown.
DESCRIPTION: Angora goats originated in the Ankara region of Turkey. They are an ancient Middle Eastern breed dating to the time of Moses. The first Angora goats in North America arrived in 1849 when Sultan Abdülmecid I presented seven Angora goats to Dr. James P. Davis. Angora goats grow long, lustrous locks of fleece up to 12 inches (30 cm) in length; this fleece is called mohair (angora fiber comes from Angora rabbits). Angora goats are generally shorn twice a year, when locks are 4 to 6 inches (10–15 cm) long. Fleece falls into one of two growth patterns: ringlets or flat.

Angoras are medium-size goats — does weigh roughly 80 to 110 pounds (36–45 kg) and bucks weigh up to 200 pounds (91 kg). Angoras have heavy, drooping ears and a straight profile. They are happy, laid-back goats that also make grand pets. They are, however, more delicate than many other breeds and require weatherproof shelter after shearing (as opposed to just covers). Angora does are excellent mothers. They are likely to produce a single kid, though twinning isn't unheard of. There are three types: modern white Angoras, modern colored Angoras, and traditional Navajo Angoras. All three are registered by separate organizations.

ARAPAWA

ARAPAWA (*arah-PAH-wah*)

TYPE: Heritage multipurpose
ORIGIN: New Zealand
COLOR: Black, brown, tan, cream, and white in varying combinations; many have badger stripes on their face
DESCRIPTION: If you're interested in protecting a breed that is close to extinction, then the Arapawa may be for you. As of 2008, there were only 500 of these goats worldwide. Extremely endangered Arapawa goats are listed as Critical on the American Livestock Breeds Conservancy Priority List (see box on page 33).

Fanciers claim that Arapawa goats are descended from Old English goats that arrived in New Zealand during eighteenth-century British colonization, with a feral population established on Arapawa Island since that time. They are handsome, medium-size goats with a fringe of shaggy hair on their hind legs and a medium-length coat; both sexes have beards. Does weigh 60 to 80 pounds (27–36 kg) and bucks weigh up to 125 pounds (57 kg).

Oh-So-Soft Cashmere Goats

Cashmere goats are a type, not a breed, although Pedigree International registers Cashmere goats. Though all goats except Angoras produce some cashmere (cashmere is their soft winter undercoat), Cashmere goats produce considerably more than the norm. The Cashmere goat standard from the Eastern Cashmere Association states in part that Cashmeres produce at least 2 ounces of processed down measuring 18.5 microns or less each year.

BOER

BOER (*BOH-er*)

TYPE: Meat

ORIGIN: South Africa

COLOR: Traditional (white with a light brown to dark red head), black traditional (white with a black head), paint (spotted), red, black. Most Boers have white facial markings.

DESCRIPTION: The Boer is the quintessential meat goat. Both sexes are large-framed and massively built. Mature bucks and wethers weigh 250 to 300 pounds (113–136 kg); does, 150 to 200 pounds (68–91 kg). They are also gentle, intelligent, easygoing goats that make wonderful harness goats as well as packgoats for use on flat to rolling terrain.

Boers have long, pendulous ears and a Roman-nosed (convex) profile; sturdy, swept-back horns; and a short coat. Many bucks have rolls of loose skin on their shoulders. Does are prolific (twins and triplets are the norm), good milkers, and excellent mothers. Boers breed year-round.

Boer and Kiko Composite Breeds

Maybe you can't decide between a Kiko and a Boer. Well, you can have the best of both breeds! The American Kiko Goat Association maintains a herd book for Genemaster goats. First-generation crosses are created by breeding full-blood Kikos to full-blood Boer goats. The best from that generation are mated with full-blood Boers to create one-quarter Kiko and three-quarter Boer offspring. These animals are crossed with first-generation Kiko and Boer crosses to create the three-eighths Kiko and five-eighths Boer Genemaster goat.

The International Kiko Goat Association registers three Boer and Kiko cross-breeds: the **BoKi** (half Kiko, half Boer), the **American MeatMaker** (half BoKi, half Kiko), and the **International Meat-Maker** (half BoKi, half Boer) goat.

GOLDEN GUERNSEY AND BRITISH GUERNSEY

TYPE: Dairy

ORIGIN: Guernsey, UK

COLOR: All shades of gold, with or without small white markings and a blaze or star on the head

DESCRIPTION: The Golden Guernsey is a unique, endangered breed that originated on the Channel Islands, off the coast of England. Golden Guernseys are recognized as a dairy breed by the British Goat Society. They are handsome, fine-boned, medium-size goats, and though they produce less milk than Nubians or the Swiss breeds, their milk is high in butterfat and proteins, making it ideal for cheese making. Both long- and short-hair coats are acceptable.

The British Guernsey is 96.8 percent or more Golden Guernsey; most resemble pure-breds to the letter. Guernsey Goats in America and British Guernseys are, at this time, registered with the British Goat Society.

KIKO (*KEE-koh*)

TYPE: Meat

ORIGIN: New Zealand

COLOR: Any

DESCRIPTION: Kikos hail from New Zealand, where their name means "meat" in Maori. Their origins date back to the early 1970s, when goat ranchers collected and bred thousands of feral goats, keeping back only the fastest-maturing, meatiest, and most disease- and parasite-resistant individuals from each generation to use as breeding stock. The ranchers provided these goats with no supplementary feed, shelter, hoof trimming, or vet care, and no assistance at kidding, so only the strong survived.

Kikos are large, muscular goats with medium-length, non-pendulous ears; a straight profile; and (especially in bucks) enormous horns. Their coat is short to medium length. Most American Kikos are white, though Kikos come in a variety of colors. Like the other meat goat breeds, Kikos make stellar harness and packgoats. Plus, their unique horns make them a standout in any crowd.

KINDER

TYPE: Dairy, meat

ORIGIN: Washington, USA

COLOR: Any

DESCRIPTION: Kinder goats are small, dual-purpose animals raised for milk and meat. They're a good choice for the homesteader with limited space. Some Kinder does give an impressive amount of milk, and butterfat content ranges from 5½ to 7½ percent.

Kinder goats have long, wide ears, resting below horizontal and extending to the end of the muzzle or beyond when held flat against the jaw line. Their facial profile is straight or dished and they have a short, fine-textured coat. Adult does are 20 to 26 inches (51–66 cm) tall; bucks, a maximum of 28 inches (71 cm). A registered Pygmy buck bred to a registered Nubian doe produces first-generation offspring; after that, Kinders must be bred to other Kinders.

Miniature Dairy Goats

The Miniature Dairy Goat Association registers Mini-Alpines, Mini-LaManchas, Mini-Nubians, Mini-Oberhaslis, Mini-Saanen/Sables, and Mini-Toggenburgs. Adult does must, however, be at least 23 inches (58 cm) tall; mature bucks, 24 inches (61 cm). Maximum heights vary by breed, but the preferred maximum height is 1 inch (2.5 cm) shorter than the minimum height for the corresponding full-size breed.

To create the F1 generation of any of the Miniature Dairy Goat Association breeds, breeders mate a Nigerian Dwarf buck to a registered, full-size doe of the desired breed; the offspring are registered as Experimentals. Through continual upgrading, breeders create registered Americans at the third generation and Purebreds at the sixth generation. The organization's goal is to produce compact, high-production does suitable for hand milking on the small-scale farm. Miniature dairy goats produce between 2 pounds (1 pint) and 10 pounds (well over 1 gallon) of milk per day.

LAMANCHA AND MINI-LAMANCHA

TYPE: Dairy
ORIGIN: California, USA
COLOR: Any
DESCRIPTION: LaManchas are sturdier than the Swiss dairy breeds and are noted for their peerless personalities, docile nature, and steady production of milk with moderately high fat levels. They have short, glossy hair and a straight profile. The breed's most unusual characteristic is its seeming lack of external ears. There are two types of LaMancha ears: gopher and elf. Gopher ears lack cartilage but have a ring of skin around the ear opening.

MINI-LAMANCHA

Elf ears are triangular, external ear flaps that are up to 1 inch (2.5 cm) long. Does may have either type, but only gopher-eared bucks can be registered.

LaManchas are the only major dairy breed developed in North America. They probably descend from short-eared Spanish goats that accompanied early Spanish padres to the West Coast. Their name may have been given to the breed when a crate of unusual, short-eared goats arrived at the Paris World's Fair for exhibition in 1904; the crate bore the inscription LaMancha, Córdoba, Spain.

Mature Mini-LaMancha does must be a minimum of 23 inches (58 cm) tall and a maximum of 27 inches (68.5 cm). Height requirements for miniature bucks are 24 inches (61 cm) minimum, 29 inches (74 cm) maximum.

MINIATURE SILKY FAINTING GOAT

TYPE: Pet
ORIGIN: Virginia, USA
COLOR: Any
DESCRIPTION: Miniature Silky goats have a straight, flowing coat that almost sweeps the ground. Their hair is lustrous, smooth, and silky to the touch, resembling that of a Silky Terrier dog. Renee Orr, of Lignum, Virginia, developed this unique pet breed by crossing two long-coated Myotonic bucks with a group of long-coated Nigerian Dwarf does with Myotonic goats in their background.

Miniature Silkies have erect ears and a dished face. Adult does may be as much as 23½ inches (59.5 cm) tall; adult bucks, 25 inches (63.5 cm). Not all Miniature Silky Fainting goats "faint," though many do.

Do Fainting Goats Really Faint?

No, they don't. They are, however, affected by a genetic disorder called myotonia congenita, an inherited, neuromuscular disorder that makes muscles unable to relax after contraction. When Myotonic goats are startled, skeletal muscles, especially in their hindquarters, contract, hold, and then slowly release. Episodes are painless and the goats remain awake until the stiffness passes.

NIGERIAN DWARF

MYOTONIC (*My-oh-TAHN-ik*)

Also called Tennessee Fainting Goats, Fainters, Texas Wooden Legs, Nervous Goats, Stiff-Legs, and Scare Goats. The Tennessee Meat Goat is a trademarked breed of large, meaty Myotonic goats.

TYPE: Pets, meat

ORIGIN: Tennessee, USA

COLOR: Although black and white is their traditional color, all colors are acceptable.

DESCRIPTION: The Myotonic goat is a rare breed that the American Livestock Breeds Conservancy has placed in the Watch category (see page 33). It is heavy-boned with massive hindquarters. Myotonic goats can vary greatly in size (anywhere from 45 to 200 pounds [20–91 kg]), be long- or short-haired and polled or horned, but all share a subdominant and recessive trait called myotonia congenita, which causes them to "faint" when startled (see the box on page 39).

Does are easy kidders and excellent mothers; twins and triplet kids are the norm. Myotonic goats don't climb or jump fences in the manner of other goats, making them a popular choice with many goat keepers; however, because they may "faint" instead of running

from predators, good fences are still necessary to keep out the bad guys. Although Myotonics make excellent pets, their propensity to stiffen and fall makes them unsuitable for carting or packing.

NIGERIAN DWARF

TYPE: Dairy

ORIGIN: Nigeria

COLOR: Any, although Pygmy-specific markings are penalized

DESCRIPTION: Nigerian Dwarfs are perfectly proportioned miniature dairy goats. Mature does are ideally 17 to 19 inches (43–48 cm) tall, though does up to 21 inches (53 cm) tall are acceptable. The ideal size of mature bucks is 19 to 21 inches (48–53 cm); bucks up to 23 inches (58 cm) tall are acceptable. The ideal weight is around 75 pounds (34 kg).

Nigerian Dwarfs have erect ears, a straight profile, and a soft coat with short- to medium-length hair. Does give up to 2 quarts (4 pounds) of 6 to 10 percent butterfat milk per day. Nigerian Dwarf kids weigh about 2 pounds (0.9 kg) at birth but they grow quickly. "Nigies" breed out of season and litters of two to four kids are the norm.

GOAT BREEDS

ALPINE

ANGORA

ARAPAWA

BILBERRY

BOER

CASHMERE

GOLDEN GUERNSEY

KINDER

LAMANCHA

MYOTONIC

NIGERIAN DWARF

NUBIAN

OBERHASLI

PYGMY

PYGORA

SAANEN

SAN CLEMENTE

SPANISH

TOGGENBERG

NIGORA

TYPE: Fiber

ORIGIN: United States

COLOR: Any

DESCRIPTION: Nigoras are a cross between Nigerians and Angoras. Nigora fiber comes in three styles: type A (Angora type), falling in long, lustrous, curly or wavy locks up to 6 inches (15 cm) in length; type B (also known as cashgora), blending Angora mohair with the soft cashmere undercoat of the Nigerian Dwarf; and type C (cashmere type), consisting of fine, nonlustrous cashmere fiber, 1 to 3 inches (2.5–7.5 cm) in length, overlaid with coarser guard hair.

Nigoras can have erect ears like Nigerians, droopy ears like Angoras, or any type in between. Their facial profile is straight to slightly dished.

F1-generation Nigora goats are created by breeding registered Nigerian Dwarf bucks to registered, full-size Angora does (the reverse breeding is acceptable, but it is risky due to potential problems at kidding time). Thereafter, any matings that result in offspring that are no more than 75 percent of one breed or no less than 25 percent of the other are accept-

NUBIAN

able. Nigora goats that are predominantly of Angora breeding are called heavy Nigoras (for their larger size and heavier fleece), Nigoras from predominantly Nigerian breeding are called light Nigoras (for their smaller size and lighter fleece production), and goats that are of roughly half-Angora and half-Nigerian breeding are known as standard Nigoras.

NUBIAN AND MINI-NUBIAN

TYPE: Dairy

ORIGIN: England, where they are called Anglo-Nubians

COLOR: Any

DESCRIPTION: Nubians are known for their long, floppy ears; short, silky hair; and beautifully Roman-nosed profile. They are elegant and graceful creatures, intelligent and inquisitive with endearing, quirky personalities. They are heavier-bodied and give milk with a higher butterfat content than do the other dairy breeds.

The English developed Anglo-Nubian goats by crossing native British does with Jumna Pura, Zaraibi, and Chitral bucks from Africa and India. A British breed registry formed in

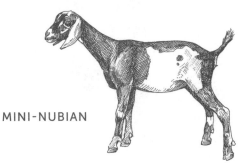

MINI-NUBIAN

1919. Goats imported by J. R. Gregg of California in 1909 and 1913 formed the nucleus of the breed in North America. The breed's major drawbacks, some say, are its unusual talkativeness and its voice, which is loud and somewhat strident — traits that make the goats less than perfect for those with close neighbors.

Mature Mini-Nubian does must be at least 23 inches (58 cm) tall and no more than 29 inches (74 cm). Mature miniature bucks must be 24 to 31 inches (61–79 cm) tall.

Vintage photograph of a happy goat owner cuddling what appears to be a Myotonic kid.

OBERHASLI AND MINI-OBERHASLI

TYPE: Dairy
ORIGIN: Switzerland
COLOR: Chamois (bay-colored, sometimes described as being colored "like the wood on the back of a violin") with black dorsal stripe, udder, belly, and lower legs; the head is nearly black with two white stripes on the sides. Black does, but not black bucks, can be registered.
DESCRIPTION: Oberhaslis are noted for sweet, tasty milk; intelligence; and peerless dispositions. They are slightly smaller than the other Swiss breeds, and they have erect ears, a straight or dished profile, and short, silky hair.

Oberhaslis originated in the canton of Bern. They came to the United States in the 1920s but were at that time considered a type of Alpine

MINI-OBERHASLI

PYGMY

goat. Esther Oman, a breeder in California, kept a group of purebred "Swiss Alpines" (Oberhaslis) descended from those early imports and almost single-handedly maintained the breed for many years. They weren't accepted by the American Dairy Goat Association as a separate breed until the 1960s.

Mature Mini-Oberhasli does must be 23 to 27 inches (58–68.5 cm) tall; mature miniature bucks, 24 to 29 inches (61–74 cm).

PYGMY GOAT

TYPE: Pet

ORIGIN: West Africa

COLOR: Pygmy colors are caramel, gray, brown and black agouti, black, and spotted

DESCRIPTION: The Pygmy is the quintessential pet goat, though it was raised for meat in its African homeland. The goats are hardy, agile, alert, animated, good-natured, and gregarious. Does give from 1 to 2 quarts (2–4 lbs) of milk (ranging from 5 to greater than 11 percent butterfat) that the National Pygmy Goat Association claims is higher in calcium, phosphorus, potassium, and iron than milk from full-size dairy breeds.

Pygmies are short, wide, compact, full-barreled, and muscular, with a proportionally greater weight and bulk than miniature dairy breeds. Does 1 year old and older must be at least 16 inches (40.5 cm) tall; the maximum height is 22¾ inches (57 cm). Bucks 1 year old and older must be at least 16 inches (40.5 cm) tall; 23⅝ inches (59.5 cm) is their upper limit. Pygmy goats have erect ears, a dished profile, and a full coat of straight, medium-length hair.

SAANEN

PYGORA

TYPE: Fiber

ORIGIN: Oregon, USA

COLOR: All Pygmy colors and their dilutions, plus white. Among breed-specific markings are a dorsal stripe, socks to the knees that can be incomplete or dilute, and a facial mask; dark animals sometimes have a white crown.

DESCRIPTION: The Pygora breed originated in the early 1970s when handspinner Katherine Jorgensen began crossing Pygmy and Angora goats in an effort to re-create the fiber she saw growing on Navajo goats living on an Arizona reservation.

Pygoras have one of three types of fiber. Type A fiber averages 6 inches (15 cm) in length and hangs in long, lustrous ringlets; it may be a single coat, but a silky guard hair is usually present. Type B fiber has characteristics of both mohair and cashmere; it's double-coated, generally curly, and averages 3 to 6 inches (7.5–15 cm) in length. Type C is very fine fiber, usually below 18.5 microns in diameter, and can be marketed as cashmere; it must be a least 1 inch long and is usually 1 to 3 inches (2.5–7.5 cm) in length.

Pygoras have medium-long, drooping ears and a straight or dished profile. Does are 22 inches (56 cm) tall on average; the minimum size is 18 inches (46 cm) at 2 years of age. Bucks average 27 inches (68.5 cm); the minimum is 23 inches (58 cm) at 2½ years of age. There is no size limit.

SAANEN (*SAH-nen* OR *SAW-nen*), MINI-SAANEN, AND SABLE

TYPE: Dairy

ORIGIN: Switzerland

COLOR: Saanens are white or cream with pink or olive-colored skin; all other colors are Sables

DESCRIPTION: Saanen goats originated in the Saanen Valley of the canton of Bern, where they were selected for milking ability, hardiness, and color. In 1893, several thousand head of Saanens were taken from the valley and dispersed throughout Europe; they came to the United States in 1904 and became the first goat breed registered in North America.

Saanens and Sables are big-boned goats with erect ears, a straight or slightly dished face, and a short coat. They are heavy milkers (some call them the Holsteins of the goat

MINI-SAANEN

world). Saanens and Sables are strong, vigorous goats with lovable personalities and an easygoing outlook on life. As is the case for the other Swiss breeds, however, they don't fare well in hot, humid parts of the world, and their light-colored skin predisposes Saanens to skin cancer. In southern climates, Sables fare better.

Adult Mini-Saanen does must be 23 to 29 inches (58–74 cm) tall. Mature miniature bucks must be 24 to 31 inches (61–79 cm) tall.

Sable or Saanen?

Saanens are solid white or cream-colored goats with light-colored skin. Sables are colored Saanens with pigmented skin (though Sables are now considered a separate breed). The first Sables in the United States arrived on the same ship as the first Saanens and have been here ever since. Sables are the result of pairing two recessive genes, one from an animal's sire and one from his or her dam. If an animal has only one of these genes, it is white or cream.

SAN CLEMENTE ISLAND

TYPE: Heritage multipurpose

ORIGIN: San Clemente Island, off the coast of southern California, USA

COLOR: Light brown to dark red or amber with markings as follows: black head with two brown stripes down the face from above or around the eyes to the muzzle; a black patch on the cheek or jaw and a small black spot on the chin; a black cape over the shoulders, up the top of the neck, and down the front legs. The underside of the neck is brown. Ears are black outside and brown inside. A black dorsal stripe runs down the back, and there is black on the hind legs and flanks.

DESCRIPTION: San Clemente Island goats are an extremely endangered breed listed as Critical on the American Livestock Breeds Conservancy Priority List. They are medium-size goats with narrow, horizontally carried ears that have a distinctive crimp in the middle of each. They are dish-faced, leggy, sleek, and deerlike.

Although a good deal of colorful folklore surrounds these goats' origin, it appears that they came to San Clemente Island in 1875, carried there by a man who claimed to have transferred them from nearby Santa Catalina Island. They thrived on San Clemente Island as feral goats until the mid-1980s, when the U.S. Navy, which took possession of the island in 1934 and maintains a naval base there, ordered their extermination or removal. The Fund for Animals saved the breed from extinction by removing more than 6,000 goats, but it soon lost track of most of them. However, the new San Clemente Island Goat Association registered 375 goats (254 does, 77 bucks, and 44 nonbreeders) by fall of 2008, and the estimated global population now stands at 425 head.

SAVANNA (*Suh-VAN-nuh*)

TYPE: Meat

ORIGIN: South Africa

COLOR: White with black skin, horns, nose, and hooves

DESCRIPTION: Savannas are big goats; they are long, broad, and muscular. These powerful white beauties make outstanding driving goats as well as sturdy packgoats for use on level to rolling terrain, where their shorter, meat-breed legs are not a disadvantage.

Savannas have long, pendulous ears; sweptback horns; and a short coat. Bucks have rolls of loose skin on their shoulders. Savanna goats resemble their South African cousin, the Boer goat, in size, shape, and disposition. However, those who breed them say they are hardier and less prone to kidding problems than are Boer goats.

FRIENDS.

TOGGENBURG

SPANISH

TYPE: Heritage multipurpose, but mostly meat
ORIGIN: United States (of Spanish stock)
COLOR: Any
DESCRIPTION: Christopher Columbus brought the first Spanish goats to Española (Haiti) on his second voyage, in 1493; when Francisco Vasquez de Coronado led the first Spanish Army into New Mexico, in 1540, goats came along as a walking food supply. Spanish colonists and their goats subsequently settled in what are now our southwestern and southeastern states. Goats that escaped or were set free to fend for themselves became the ancestors of today's Spanish goats. Because Spanish goats are a landrace breed (one that adapted to the region where it lives) and a lot of outcrossing to other breeds occurred during the twentieth century, genetic diversity within the breed is expected.

Individuals weigh between 50 and 200 pounds (23 and 91 kg); strains selected for meat-making capabilities are made up of larger animals. All Spanish goats have an inherent wariness, strong teeth and legs, almost unimaginable hardiness, and the ability to take care of themselves. Because they are such easy keepers and because they kid without assistance, thousands upon thousands of tough, rangy Spanish does throughout the country are being bred to Boer and Kiko bucks (instead of to other Spanish bucks) to produce meaty slaughter kids. This crossbreeding means that the breed is quickly disappearing. Only an estimated 8,000 reasonably pure Spanish goats remain. The breed is listed as Watch on the American Livestock Breeds Conservancy Priority List (see box on page 33); additional conservation breeders are needed.

TOGGENBURG AND MINI-TOGGENBURG

TYPE: Dairy
ORIGIN: Switzerland
COLOR: Base colors range from light fawn to dark chocolate, but all Toggenburgs have the same markings: white ears with a dark spot in the middle of each ear; two white stripes from above each eye to their muzzle; hind legs that are white from hocks to hooves and forelegs that are white from knees downward (a dark band below each knee is acceptable); a white triangle on both sides of their tail; and a white spot at the root of their wattles, or in that area

MINI-TOGGENBURG

AUSTRALIAN MINIATURE

if no wattles are present. Cream markings are also acceptable.

DESCRIPTION: Toggenburgs are outstanding milkers; a Toggenburg doe, GCH Western-Acres Zephyr Rosemary, holds the *Guinness Book of World Records* title for giving 9,110 pounds of milk — amounting to almost 1,140 gallons in 365 days. Many Toggenburg does "milk through" without rebreeding each year; lactations of 18 to 20 months are common.

The Swiss developed Toggenburg goats about 300 years ago in the Toggenburg Valley, in the northeast part of the country. The breed's supporters call it the oldest and purest of the Swiss breeds. "Toggs" came to America in 1883 and were subsequently the most popular dairy goat breed to import.

Toggenburgs have a medium-length coat; a straight or dished face; alert, upright ears; and a high, globular udder. Mature Toggenburg bucks stand 34 to 38 inches (86–96.5 cm) tall and weigh 150 to 200 pounds (68–91 kg); does measure 30 to 32 inches (76–81 cm), weighing 125 pounds (57 kg) or more.

Adult Mini-Toggenburg does must be 23 to 25 inches (58–63.5 cm). Adult miniature bucks must be 24 to 27 inches (61–68.5 cm).

Breeds across the Sea

The following breeds aren't available in North America, but maybe they should be. Strict regulations forbid the import of live goats from abroad, but importing frozen embryos and semen through firms that specialize in livestock genetics is usually permitted. Perhaps you'd like to import frozen genetics and create an American version of one of these interesting breeds: that's how Golden Guernseys, Boers, and Kikos got here.

AUSTRALIAN MINIATURE GOATS

An Australian Miniature goat's conformation is similar to that of larger breeds, with all parts of the body in balanced proportion to its size. Miniature-goat fanciers, take note!

Two organizations register Australian Miniature goats: the Miniature Goat Breeders Association of Australia and the Australian Miniature Goat Association (see Resources). Maximum height for adult Australian Miniature goats registered with the Miniature Goat Breeders Association is 21 inches (53 cm); purebred adults registered with the Australian

BAGOT

Miniature Goat Association must be no more than 20 inches (51 cm).

The Australian Miniature Goat Association recognizes three breed types based on coat and ear length: Minikin (a short cashmere coat that is shed and any ear type that is shorter than muzzle-length), Sheltie (long, nonshedding coat and any ear type), and Nuwby (short, shedding cashmere coat and longer-than-muzzle-length ears).

The Miniature Goat Breeders Association of Australia recognizes four types based on ear style: Elf (short ears), Pixie (upright ears), Munchkin (folded ears), and Nuwby (pendulous ears).

BAGOT GOAT

The Bagot's history harkens back to 1387, when King Richard III supposedly gave its ancestors to Sir John Bagot, who incorporated them into his family coat of arms. They have been documented running semi-wild at the Bagot family home, Blithfield Hall in Staffordshire, for at least 600 years. Bagot goats are listed on the British Rare Breed Survival Trust Watchlist under category 2: Endangered (see the box below). They need additional dedicated conservators to survive.

Bagot goats are small to medium size. Both sexes have large curving horns, long hair, and a distinctive color pattern: black forequarters with white rear quarters.

British Rare Breeds Survival Trust (RBST) Watchlist

The Watchlist contains all native breeds of cattle, goats, horses, pigs, poultry, and sheep. Criteria for each category are as follows.

Critical. Fewer than 100 individuals remain; radius of geographic concentration is less than 12.5 km (7.8 mi).

Endangered. Fewer than 200 individuals remain; radius of geographic concentration is 12.5 to 15 km (7.8–9 mi).

Vulnerable. Fewer than 300 individuals remain; radius of geographic concentration is 15 to 17.5 km (9–10.9 mi).

At Risk. Fewer than 500 individuals remain; radius of geographic concentration is 17.5 to 20 km (10.9–12.4 mi).

Minority. Fewer than 1,000 individuals remain.

Other native breeds. Fewer than 1,000 individuals remain.

BRITISH ALPINE

The British Alpine is a superb dairy goat. It was developed in England during the early twentieth century by combining Alpine, Toggenburg, Nubian, and native British goat genetics. The founder of the breed was Sedgemere Faith, who imported a black-and-white Swiss-marked doe from the Paris Zoo in 1903. The doe was the first British goat officially recorded as giving a gallon of milk in 24 hours. The first British Alpine herd was established in 1911.

British Alpines are big, active dairy goats that always sport black-with-white Swiss markings. They are strong and rangy, well suited to browsing, and have a reputation for pleasant-tasting milk. British Alpines tend to milk through, requiring breeding only every other year.

BRITISH SAANEN

British Saanens are big white dairy goats developed in the United Kingdom by combining Saanen, mixed Swiss, and native British goat genetics. They have longer legs than purebred Saanens and are heavier. They have a calm nature and offer high milk yields and long lactations. According to the British Goat Society, these qualities make it a popular breed where large groups of dairy goats are housed together. For these reasons, British Saanens are kept by some of the largest goat farms in the United Kingdom.

Progeny of registered purebred Saanens and British Saanens can still be registered as British Saanens.

BRITISH TOGGENBURG

These are outstanding dairy types that tend to milk through. British Toggenburgs have lovely temperaments and, like British Alpines, they thrive on forage-based diets. The colors and markings on British Toggenburgs are similar to those of pure Toggenburgs, but British individuals are larger, have shorter coats, and some say they milk much better.

British breeders developed the British Toggenburg by mating imported Toggenburg bucks to does of mixed Swiss backgrounds. The British Goat Society opened a herd book for British Toggenburgs in 1925 and a British Toggenburg Register in 1936.

ENGLISH GOAT

The English goat was historically Britain's household dairy goat until the early twentieth century, when they were replaced by bigger, higher-yielding imports and their crosses. Today's English goat breed standard is the same as the one published in 1919.

The English goat is a medium-size animal that is sturdy and deep. Both sexes are horned and bearded. Does have a short, dense outer coat with longer fringe along their back and flanks, and thicker hair growth on their legs; bucks have longer hair throughout but especially on their back, neck, chest, and thighs. English goats are brown or gray, with a dark eel stripe along the spine and dark markings on the legs, neck, and flanks.

The English goat is another fine browser. It is tractable and docile and generally milks through for two years.

This photo was taken on Okinawa during World War II. What are the soldiers saying to the doe and her tiny kid? We'd like to know!

OLD ENGLISH GOAT

The Old English goat, like the English goat, was the typical cottager's goat for centuries. Today, however, there are fewer than 50 licensed Old English goat breeders in the United Kingdom and only about 100 registered goats. British feral goats, being descendants of Old English goats, are accepted in the breed's herd book. This breed, too, needs additional conservators to survive.

The Old English goat is a small, cobby, hardy, and thrifty goat noted for long lactations. It has a large rumen adapted for dealing with coarse browse. Colors include brown and gray, with darker legs, and some have white patches. The coat is thick; the beard, pantaloons, and other hairy trimmings are the norm.

Part 2

Playing and Working with Your Goats

CHAPTER 4

Training and Tricks

*In time, even a goat
can be taught to dance.*

~ Yiddish proverb

For many people, one main appeal of goats is their quirky, winsome personalities. When you get a goat, you may find yourself looking for more ways to spend time with her beyond the usual chores. Goats are immensely fun to "play" with, and one such way to do this is by training your goat to perform tasks and tricks. You can train your goat to follow you with a loose lead, pick up her feet for hoof trimming, and perform a fun trick like doing a spin on command or "laughing" when you tell a joke.

Goats are smart and they aim to please, so training them is a fun and easy task. Although most any training system based on positive reinforcement works well, clicker training is the perfect way to train goats. Yes, clicker training, the same technique used to train dogs, horses, and sea mammals such as dolphins.

Clickers have proved to be a quicker and more effective way to train than using lan-

guage alone. Unlike words, a clicker enables you to pinpoint the exact moment a correct behavior is achieved. Consider a study conducted by Lindsay A. Wood, MA, CTC, and reported in her Hunter College master's thesis, "Clicker Bridging Stimulus Efficacy." In this study, the first of its kind, two groups of untrained dogs were taught to perform the same task: walking across a small room and touching a stationary target. The only difference in their training was the bridging signal used: a click or the spoken word *good*. It took the clicker-trained dogs one-third less time to learn the task than the dogs that received the oral signal.

Clicker training is fun and so rewarding, but unfortunately many people interested in the technique never learn how to do it because they don't understand how the system works. Instruction materials are often couched in confusing jargon and are difficult for beginners to follow, especially factoring in the need to translate dog- or horse-oriented techniques into something usable for training goats. That's why we'll stick to everyday language in this book.

Clicker-Training Prep

You won't have to buy a lot of gear or make elaborate arrangements to clicker-train your goat. Here are a few things you will need to begin.

A clicker. Buy a clicker at a pet store, on eBay, or from clicker-training pioneer Karen Pryor's clicker-training store (see Resources).

A target. A target can be anything from an empty soda pop bottle to your hand. I like the handheld marine-float type used by clicker-training guru Shawna Karrasch (see Resources). Buy one from her store, make your own (see page 67), or improvise with whatever is at hand.

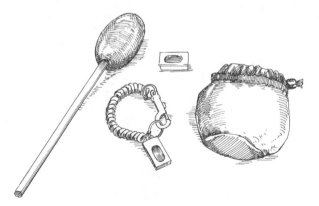

Basic training equipment comprises a target like the marine-float model to the left; a clicker, the handheld kind or a clicker attached to a stretchy cable you can wear like a bracelet or clip to your belt; and a goody bag.

Rewards. In most cases this means food. Keep in mind that each reward is a teensy tidbit; you're rewarding, not feeding, your goat. Rewards should be something your goat loves to eat, broken into pieces no bigger than a raisin.

A goody bag. Rewards should be stowed out of sight unless you're using food as a lure. They can be stored in a loose-fitting pocket or kept in a separate receptacle. Carpenter's aprons, the fabric kind lumberyards hand out for free, make great goody bags, as do horse- and dog-training reward bags designed to clip on your pocket or belt. Fanny packs make nice, inexpensive goody bags, too.

A quiet, out-of-the-way spot to train your goat. Reduce distractions so that your goat can focus on the task at hand.

Real Rewards

Before you start your clicker-training project, discover what types of goodies your goat likes best. Collect a selection of items, spread them out, and observe what your goat seems to favor.

These are a few things to try.

- Dry breakfast cereal (Cheerios are perfect)
- Animal crackers, broken into smaller pieces
- Raisins or another dried fruit, chopped into tiny bits
- Pelleted goat or horse feed, especially if it's not on your goat's everyday menu
- Popped popcorn
- Packaged croutons (broken to size if needed)
- Peanut halves (whole peanuts in the shell are jackpot fodder for many goats)

- Sunflower seeds, shelled or otherwise
- Pieces of tortilla chips or pretzels
- Chopped fresh fruit and vegetables such as apples, pineapple, pears, carrots, celery, and kohlrabi (this is messy; line your goody pouch with a plastic bag)
- Premium-grade vegetarian dog kibble (don't feed meat-based products to goats)

Keep in mind that some goats have never been given "people food" and may be reluctant to try it. Allow the reluctant goat to watch other goats eating treats (goat see, goat do) or gently open his mouth, insert a goody, and let him taste it. He may spew out the same item several times before deciding it's really very good. Be patient. And many goats work well for nonfood rewards. For these goats, lavish praise or a scratch behind the ears may be enough.

If there is one item he loves above the others, save that treat for "jackpots," which are super-special rewards for outstanding behavior. Jackpots can be a special treat or a larger-than-normal amount of his usual training snacks. Don't hand them out indiscriminately — they'll lose their value as a reward.

The Ten Commandments of Clicking

Clicker training is enjoyable and easy, but there are a few important things you should know before you start.

1. Keep sessions short and make them fun.
Five-minute sessions two or three times each day are ideal (and more productive than a single 1-hour session).

2. Click at precisely the right moment.
Picture this: you're trimming your goat's hoof and he holds up his leg like a prince; you put the leg down, then click and reward. You think you're clicking for holding his foot up, he thinks you're clicking because he put it down. Which action is he likely to do the next time you click? Click at the precise moment a desired behavior occurs, not seconds before or after.

> ## Debunking the Myths
>
> Some goat owners won't try clicker training because they think that once they do, their goats will mob them for food. Quite the reverse is true. Once a goat realizes that food appears only after he's heard a "Well done!" click, he'll stop expecting handouts all the time. Once you begin clicker training, however, *never* indiscriminately feed treats from your hand. You can still treat your goat to random goodies, but place them in a feed pan or on the ground and step away. Or ask him to work for his treats.
>
> Other misconceptions are that once you've clicker-trained a goat you'll always need a clicker to make him work, and that the goat you train will work only for food. Clicks and rewards are used to teach *new* behaviors; once a goat learns them, the clicker and food rewards are phased out and simple praise can take their place.

LEARN THE LINGO

You needn't understand the bewildering array of terms encountered in many clicker-training books to learn how to successfully clicker-train your goat. These, however, are a few terms you should know.

BEHAVIOR. Any action your goat takes.

BRIDGING STIMULUS. Something (like a click from your clicker) that tells your goat "That's what I want you to do!" The click bridges the time between his correct response and the delivery of a reward.

CLICKER. A small, handheld box usually made of plastic containing a metal tongue that emits a sharp click when pressed and released.

CONDITIONED REINFORCER. The click of a clicker, after being repeatedly associated with a food reward, is a conditioned reinforcer.

CUE. Something that elicits a behavior. Good cues for goats are words and hand signals.

EVENT MARKER. Similar to a bridging stimulus, it is a signal used to tell your goat he performed a task correctly. A click from your clicker is an event marker.

FREE SHAPING. To train using free shaping, wait until your goat performs a desired behavior on his own. The instant he does, click and reward him. For example, to teach him to wag his tail, wait until he does it of his own volition, then click and reward.

JACKPOT. An extra-stupendous reward for outstanding behavior.

LURING. A method of guiding your goat through a behavior. For example, a food lure can be used to entice your goat to hop onto a box. Lures are usually food rewards, but they can also be a target or anything else your goat will follow.

MOLDING. To train by molding, place your goat or one of his body parts in the desired position and then click and reward. An example: place one forefoot on a pedestal, then click and reward.

POSITIVE REINFORCEMENT. Something your goat will work for (usually a food reward) that is given to strengthen a particular behavior.

REWARD/REINFORCER. Anything your goat will work to obtain. Effective rewards for goats are usually tiny tidbits of food, but anything your goat enjoys (having his back scratched, a kiss on his nose) will work as well.

TARGET. Something you teach your goat to touch with his nose. Targets are often used as lures to shape behavior.

TIMING. The timing of a click from your clicker. The click should occur at the exact instant a desired behavior occurs.

3. Click once per correct behavior. If your goat does something especially well, treat him to a jackpot reward but don't reel off a series of clicks.

4. While training, click for both voluntary and accidental movements toward your goal. This sets up your goat for success. Hold the target where his nose will bump it, coax or lure him into a position you want, or briefly place him there, but don't pull, push, or hold him in place; he needs to think he earned the reward of his own accord.

5. At first, don't hold out for perfect behavior. Once your goat understands what he needs to do to earn a reward, then you can be picky, but in the early stages, click and reward for effort.

6. Correct bad behavior by clicking for good behavior. Mugging for food is a classic example: if your goat nudges and nibbles the reward bag begging for food, ignore him. Look away or turn your back and count to five. Then give him a chance to *earn* his treat by targeting or performing some other simple task. He'll quickly learn that mugging is counterproductive.

7. By the same token, don't reinforce undesirable behavior. If your goat does something wrong or engages in spontaneous behavior during training (such as targeting when you didn't ask him to), stop what you're doing and count to five to let him know you're not going to click and reward that action. Of course, it's not the counting itself that's critical — the goat doesn't know that's why you're counting to five. The point is to allow sufficient time to pass so that he doesn't associate any rewards with that action.

8. Once you're sure your goat understands a behavior, don't click for every correct response. Start clicking every second or third correct response, then stretch out the clicks until you're clicking correct responses at irregular intervals. Although it seems as though the goat would find this discouraging, quite the opposite is true — he'll wonder which response will be the magic act that earns the reward, and he'll try all the harder to get it.

9. For some goats, you may need to break down each process into many small, rewardable steps before they understand what you want them to do. Take your goat's history into consideration when planning training goals. A goat that hasn't been handled very much won't respond in the same manner as your home-raised bottle baby.

10. If you get mad or frustrated, stop. Take a deep breath, then ask your goat to perform a simple behavior you know he can do. Click and reward and then quit for the day. You don't want your goat to link anger or tears with clicker training. Clicker training is meant to be fun.

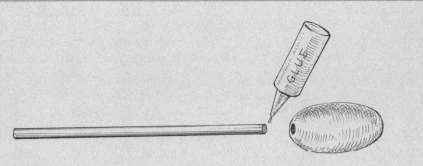

Making marine-float targets is a snap: glue a length of dowel through a float, and that's it!

MAKE YOUR OWN TARGETS

I love the type of marine-float targets that Shawna Karrasch sells on her On-Target Training website (see Resources). However, I make my own. Making them is the essence of simplicity: buy marine floats at any store that sells marine supplies (they come two to a package) and a half-inch dowel at a hardware or lumber store. Cut the dowel to length, squirt a smear of Super Glue (or its equivalent) just inside both ends of the predrilled hole in the marine float, add the handle, and there you are! Make two: a short one for close-up work and a longer one for leading; the long one can be shortened later if it's too unwieldy. (The ones I use have a 10-inch [25 cm] handle, made using a 15-inch [38 cm] piece of dowel, and an 18-inch [46 cm] handle, made from a 22-inch [56 cm] dowel.)

First Things First: Targeting

By touching his nose to a target, your goat learns to perform a task in order to earn a reward. Simple targeting leads to more-complex maneuvers such as being led quietly at your side, hopping onto a fitting stand when asked, and learning a host of neat tricks. The most satisfying part of clicker training is seeing the lights go on the first time your goat realizes what you want him to do. It's magic — and you make it happen!

First, teach your goat that a click means food. Click, then quickly hand him a tidbit of food. Goats are smart; they quickly connect the click with the treat, so don't carry this to extremes. And don't try to hide the goody bag. Show it to your goat, jiggle it, let him sniff it. He'll probably nudge and nibble the bag — or you. As long as you are safe (standing on his back legs and pawing your chest would be an exception), try to ignore unwanted behaviors at this early stage.

Begin target training by placing the target near the goat's nose.

Targeting, Phase One

Begin by teaching your goat to touch the target. Stand at his head with the target and the clicker in one hand (this is easier with a marine-float target [see box on page 67] than with something bulky) and your other hand free to quickly dish out rewards.

1. Hold the target close to your goat's nose. He'll probably reach out to see what it is. The moment his nose brushes the target, click, reward with food, and pile on the praise.

2. Keep the target close to his nose; if he doesn't immediately touch it again, position it where he'll bump it when he moves his head. When he does, click, reward, and praise.

3. Repeat this sequence until that magical moment when your goat touches the target on his own. Jackpot! Haul out the tastiest items in your goody bag and celebrate. The most difficult part of training is over.

Targeting, Phase Two

Once your goat understands that touching the target elicits a reward, teach him to follow it with his nose.

1. First move the target around within arm's length so he can touch it without shifting his feet.

2. Hold it up high so he has to stretch to touch the target and low near the ground so he lowers his head. Click and reward each time he touches the target. Then take a step back, so he has to step forward to reach it.

3. Keep moving around in a small area until he understands that following the target is your goal.

Targeting, Phase Three

When you're positive he understands the concept of targeting, and only when you're sure he'll do it 100 percent of the time, move into the last phase of target training: adding cues. Cues are words that request specific behaviors.

1. Add a spoken cue as you offer the target ("touch" and "target" are logical choices).

2. Once he understands, **reward *only* when you ask him,** by way of the cue word, to perform (ignore spontaneous touches); then you're set to start shaping behaviors. The cue becomes a stimulus for the behavior itself: the cue is followed by the behavior, and the behavior is reinforced with a click and a reward.

Once he's learned a behavior and responds to the appropriate cue, you can gradually phase out the clicker and food rewards if you like. Ask for the behavior (for example, "hoof") and reward with a "Good boy!" and a scratch behind the ears. For most goats, that's enough.

Lead Away!

When your goat reliably follows a target and understands his verbal cue, teach him to walk on a loose lead at your side. Using your long-handled marine-float target and wearing your reward pouch on your right side, stand on the left side of your haltered or collared goat. Face forward with your shoulder even with the middle of your goat's neck. The float part of the target should face away from the goat.

1. Place the float about 18 inches (46 cm) in front of your goat's muzzle and give your verbal cue (for instance, "touch").

2. When he steps forward and touches the target, click, reward, and lavish him with praise. Repeat this until he understands what you want him to do.

3. Hold the target in front of him and click as he takes a step.

4. Click for every two steps, then three. Before long, he'll be following the target with ease and you can increase the distance he travels before clicking and rewarding.

5. Introduce a new cue as he starts to walk forward (such as "walk"). When he understands, put away the target. Now you have a happy goat walking calmly at your side.

Using the same technique, you can teach your goat to hop onto a fitting stand, pull a wagon, or perform tricks such as standing with his front feet on a pedestal. When you want to teach him something new, just bring out the target and begin.

Teach a goat to lead by first putting the target about 18 inches (46 cm) in front of his nose and saying "touch."

The Wolf and the Seven Little Goats
(by the Brothers Grimm)

THERE WAS ONCE an old goat who had seven little ones, and she loved them as much as any mother could love her children.

One day she wished to go into the forest and get food for them, so she gathered them around her and said, "Dear Children, I am going out into the wood. Don't open the door while I am away, for if the wolf should get into our hut, he would eat you up, even unto the very hairs; you may easily know him by his rough voice and his large black feet."

"Dear Mother," said the young kids, "we will be very careful to keep out the wolf, don't worry." So the old goat started on her way.

She had not been away long, when there came a knock at the door, and a voice cried, "Open the door, my dear children, I have brought something nice for each of you."

But the young kids knew by the rough voice that it was the old wolf, and not their mother; so the eldest said: "We shall not open the door, you are not our mother; she has a soft and gentle voice, and your voice is rough. You are only a wolf."

Then the wolf ran away to a shop and bought a great stick of white chalk, which he ate to make his voice soft. After he had eaten it, he went back to the goats' cottage and knocked again on the door and said, in a soft voice, which the little kids thought was their mother's: "Open the door for me now, dear children; I am your mother, and I have something nice for each of you."

But the wolf put his foot on the window sill as he spoke, and looked into the room. The young kids saw it, and one of them said, "No! We shall not open the door, our mother has no black feet like that; go away, you are the wolf."

So the wolf went away again to a baker's and said, "Baker, I have crushed my foot. Please wrap it in dough; that it will soon cure it." As soon as the baker had done this, he went off to the miller and asked him to cover it with flour. The miller was too frightened to say no, so he floured the wolf's foot and sent him away. Such is the way of the world.

For the third time the wolf came to the house door and said: "Open the door, dear children, it is your mother this time; she has brought you something from the forest."

"Show us your feet," said the little kids; "then we shall know if you are really our mother." The wolf placed his white foot on the window, and when they saw that it was white, they believed all that he said was true, so they opened the door; but as soon as he came into the house they saw that he was the wolf and ran to hide themselves.

One hid under the table, another in the bed, the third in the oven, the fourth in the kitchen, the fifth in the cupboard, the sixth under the washtub, and the seventh in the clock. But the wolf found six and gobbled them up one after the other, all but the youngest, who was hidden in the clock.

Then the wolf went out quickly and laid himself down in the green meadow under a tree and fell fast asleep.

Not long after, the old goat came home from the forest. Ah! What a sight it was for her. The house door wide open; table, chairs, and stools upset; the washtub broken into pieces; the sheet and pillows dragged from the bed! She looked for her children, but not one could she find. At last she heard a little voice cry: "Dear Mother, here I am, shut up in the clock case." The old goat helped the kid out and then listened while she told how the wolf had gotten in the house and eaten up all of her brothers and sisters. We can guess how the poor mother mourned and wept for her children. At last she went out and the little kid followed her. As they crossed the meadow, they saw the wolf lying under a tree and snoring so loud that the ground trembled.

The goat looked at him on all sides and saw a movement as if something were alive in his stomach. "Ah!" she thought, "if he only swallowed my dear children, they must be alive!" So she sent the little kid into the house for a pair of scissors, a needle, and some thread and very quickly began to cut open the wolf. She had barely made one cut, when a little kid stretched out his head, and then a second, and then a third sprang out as she cut farther, until the whole six were safe and alive, jumping around their mother for joy. The greedy wolf had swallowed them whole and so had not hurt them.

Then their mother said to them: "Go and get me some large pebbles from the brook, so that we may fill the wolf's stomach while he sleeps." The seven young kids started off to the brook in great haste and brought back as many large stones as they could carry. With these they filled the stomach of the wolf; then the old goat sewed it up again so gently and quietly that the wolf neither awoke nor moved.

As soon, however, as he had his sleep out, he awoke, and stretching his legs out felt himself very heavy, and the great stones in his stomach made him feel so thirsty that he got up and went to the brook to drink. As he trotted along, the stones rattled and knocked one against the other and against his sides in a most strange manner. Then he cried out,

"What a rattle and rumble,
They cannot be bones,
Of those nice little kids,
For they feel like stones."

When he came to the brook and stooped over to drink, the weight of the stones in his stomach tipped him over, so that he fell in and was drowned.

When the little kids and their mother heard the splash, they ran over toward the brook and saw what happened. Then they danced for joy around their mother, crying out: "The wolf is dead, the wolf is dead!" And this was the end of the greedy wolf.

— *Everychild's Series: The Fairy Book*,
Kate Forrest Oswell

Trim Hooves with Ease

Few things are more frustrating than wrestling with a goat who doesn't want his hooves trimmed. Fortunately, you can clicker-train him to stand quietly for the process.

Begin with your goat standing on the ground or on a fitting stand. If he kicks or struggles when you touch his hooves, run your hand down each leg, clicking and rewarding until you reach a spot where he starts to object. When you reach the tough spot, proceed more slowly, clicking and rewarding positive responses until he's okay with your touching his hooves. Then follow these steps to teach your goat to lift his leg on command.

Nannyberries

GOAT TREATS

Goats, just like most people, love treats. However, too many goodies fed at one time easily upset a goat's digestive system. The trick to treating your goats is handing out yummies in moderation. Here are some recipes to try.

Bon Bon's Yummies

1 cup flour

1 cup sugar

½ cup of tasty tidbits such as finely diced raw apple, lightly crushed peanuts, grated carrot, raisins or other dried fruit

½ cup molasses

½ cup vegetable or corn oil

Preheat the oven to 350°F. Combine flour, sugar, and crushed or grated items, then add the molasses and oil. Stir until well blended. The mix should be sticky: if it's runny, add a little flour; if it's too thick, add a dab more molasses. Divide into small balls, place on greased cookie sheet, and bake for 10 to 15 minutes, or until crisp. Cool in the fridge if you're pressed for time.

Variation: Replace up to ¼ cup of the tidbits with crushed peppermints.

Goatmeal Apple-Carrot Balls

1 cup finely diced raw apple

1 cup shredded carrots

½ cup uncooked oatmeal

½ cup flour

½ teaspoon brown sugar

2 teaspoons molasses

⅓ cup water

Preheat oven to 350°F. Combine the ingredients in the order they are listed. Stir until well blended. Divide into small balls, place on greased cookie sheet, and bake for 10 minutes, or until crisp. Cool in the fridge.

1. Lean into your goat to shift his weight to his opposite leg; when he moves over, click and reward.

2. Run your hand down his leg, grasp his hoof, and pick it up. Click while his foot is coming up, not as he shakes his leg or slams down his hoof again. If he starts to shake his leg or his hoof starts down before you have time to click, count to five and do nothing, so that you won't reinforce the behavior, then try again.

3. When he lets you pick up his foot and holds it up for even a second, **click** and **reward** (still holding his hoof off the ground), then set it back down again.

4. Work toward three or four minutes of patient standing, slowly increasing the amount of time before he earns each reward.

5. When you're sure your goat has learned the behavior, **introduce a verbal cue**, such as "hoof" or "pick it up." Click and reward when he performs to the cue on command.

Bad Behavior Can Have a Sour Taste

Clicker trainers don't punish bad behavior; they ignore it until it goes away. This works with most species, but it's unwise to ignore a 200-pound (91 kg) wether who places his forefeet on your shoulders or bucket-dives at feeding time. People can get hurt.

A good way to put the kibosh on this behavior without betraying the ethics of clicker training is to discourage the goat in a manner he doesn't associate with punishment from you. Enter the ReaLemon, a neat, palm-sized plastic "lemon" containing lemon juice that, when squeezed, propels a stream of juice for up to 5 feet (1.5 m). To use this as a deterrent, go about your business while concealing the lemon in either hand. When your goat jumps up, don't fuss or yell. "Ignore" him while covertly spraying a strong stream of lemon juice at his mouth (be careful not to shoot it toward his eyes). As his feet hit the ground, click and reward for desirable behavior. It works!

After your goat has learned basic targeting behavior, consider teaching him to target on your hand. Once he does, it's easy to covertly lead him through intricate maneuvers, and your target is always there when you need it!

Not-So-Tricky Tricks

Shaping is the act of breaking down behaviors into a series of easy steps. Through shaping, you can teach your goat to perform any trick he's physically capable of doing, including behaviors as simple as shaking hands and as complex as standing on a pedestal and waving a flag or running agility courses. No matter the level of training, keep these points in mind.

- **All goats are different.** There is no set way to teach tricks that will work for every goat. It's up to you to discover what works best for your particular goat.
- **When you begin training a new trick, keep things simple.** Start with an easy behavior your goat already knows how to do, then ease into the new trick, building it piece by piece.
- **Always provide clear, consistent, and timely feedback;** your goat will continually look to you for cues and guidance.
- **Always train the behavior (trick) you want your goat to perform** and then attach your cue or command, not the other way around.
- **No goat is too young or too old to learn tricks.** If your goat isn't restricted by physical disabilities or health problems, you can start right now!

The following are some sample tricks to illustrate the training process. Goats can jump through hoops, bow, shake their heads yes and no, walk on their hind legs, fetch, give kisses, and so much more. Break down the training process into manageable segments, be patient, and have fun. The sky's the limit when clicker training goats!

The Talker

Wait until your goat vocalizes. (Standing 8 to 10 feet [2.5–3 m] away from a fence with you and your goat on opposite sides will often elicit a vocal response.) As he calls out, click and reward. Most goats (especially naturally noisy breeds such as Nubians and Boers) learn this trick with ease.

The Laugher

Most goats flehmen (curl up and back their upper lip) when they smell something that has a new or sharp aroma. It makes them look as if they're laughing (see page 11).

To train them to do this, begin by holding something strong-smelling (perhaps a scrap of cloth soaked in ammonia or a tiny bottle of essential oil) close to your goat's nose. When he flehmens, click and reward. When he flehmens reliably, introduce a cue (pointing at him is a good one). Then, to show off his trick, you can tell a joke, then point, and he'll laugh. It's an impressive trick — and incredibly easy to teach!

Teaching a goat to laugh or smile is ultra-easy to do.

The Greeter

To teach your goat to shake hands, stand facing him.

1. Bend over and pick up his hoof by pulling it toward you, hold it for a second, then click and reward.

2. Ask him to move his hoof by tapping his leg with your fingers. Even a slight movement of the hoof is worthy of a click and reward. The timing of the click is important; click just as your goat's foot shows a hint of movement.

3. Point at his leg without tapping. Continue clicking and rewarding until your goat offers his hoof each time you extend your hand.

4. Add a cue. I use "shake," but you can use whatever word you like. Say "shake" a split second before lowering your hand toward your goat's hoof: another neat, easy-to-teach trick!

Nannyberries

TEN GOAT BEERS AND WINES

1. Big Sky Scape Goat Pale Ale
   (*Missoula, Montana*)
2. Bored Doe Wine
   (*Suider-Paarl, South Africa*)
3. Gilded Otter Billy Goat Bock
   (*New Paltz, New York*)
4. Goat's Breath Bock Ale
   (*O'Fallon, Missouri*)
5. Goats do Roam Rosé
   (*Suider-Paarl, South Africa*)
6. Grizzly Paw Randy Goat Pilsner
   (*Canmore, Alberta, Canada*)
7. Heidrun Sparkling Mead
   (*Arcata, California*)
8. Horny Goat Beer
   (*Lacrosse, Wisconsin*)
9. Kozel Premium Lager
   (*Velké Popovice, Czech Republic*)
10. Montgomery Goat Hill Pale Ale
    (*Montgomery, Alabama*)

"Shake" is another easy trick.

The Spinner

With your goat standing in front of you, hold your long-handled float target at his eye level.

1. Move the target just to the left of his nose. At the first sign of movement toward the target, click and reward.

2. Repeat this sequence until your goat touches the target each time. If he tries to back away rather than turn toward the target, move your training session to a fence corner or against a wall to limit backward movement.

3. Move the target from your goat's nose around to the side of his body. He should follow the target. Click and reward when he's turned a quarter circle. Encourage him to turn another quarter circle; click and reward again; then continue practicing the quarter turn until you think your goat is ready to try a half spin.

4. Move the target 180 degrees around your goat (half a full circle), traveling to the right. If your goat follows the target, he'll end up facing the opposite direction from where he started. Continue to click and reward him.

5. Once your goat is comfortable with 180-degree turns, it's simple to **train him to do a full spin.** Slowly guide your target around your goat in a full 360-degree rotation. If he doesn't follow the stick for the full circle, practice the half turn again.

6. Continue to click and reward until he completes a full circle.

7. Once you've successfully trained the spin in its basic form, **ask for two or three full circles** and click and reward only the faster, tighter spins.

8. When you are certain your goat will follow the target, **add a cue for the behavior.** I use "spin," but you could use a phrase such as "chase your tail." Say your cue just before presenting the target to your goat.

9. Phase out the handheld target and ask your goat to target on your hand. When he quickly responds to your verbal "spin" cue without the need for a handheld target, phase out the clicker and the treats.

The Weaver

This is a slick trick for Pygmy goats and Nigerian Dwarfs (it's harder for the larger breeds to pass through a person's legs).

The trick is to teach your goat to pass between your legs with each step you take. You can teach it using a target or a food lure, though let's assume you're going to be using food.

To begin, place your goat on your left side, at or about what would be the heel position if he were a dog.

1. Holding a reward in your right hand, **take a big step forward with your right foot.** There should be a large gap between your right leg and your left leg.

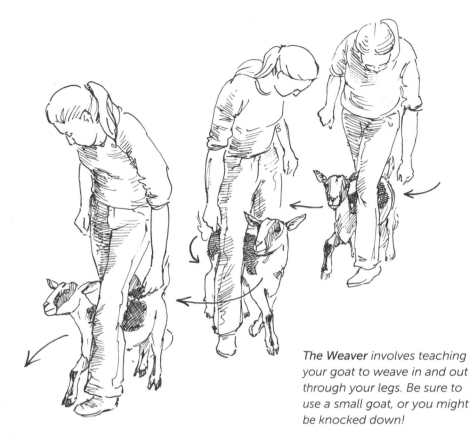

The Weaver involves teaching your goat to weave in and out through your legs. Be sure to use a small goat, or you might be knocked down!

2. Drop your hand behind your right knee and **show the food to your goat.**

3. As soon as he starts to move and follow your right hand, **click** (you are stationary for this step). Quickly lead the goat through your legs to the front of your right knee, and reward.

4. As your goat accepts the reward, **place a food reward in your left hand** and take a large step with your left leg.

5. Drop your left hand behind your left knee, **lead the goat through your legs,** and

click. Lead him to the front of your left knee, then reward.

6. Take a step with your right foot, drop your right hand behind your right knee, and **repeat the process,** leading the goat through your legs.

At first, click and reward every time the goat passes behind each of your legs. Once he's mastered that, reward for every complete weave. Then give the treat for every two weaves through your legs. Add a cue word (say "weave" or "pass"). Have fun with this cool trick!

Everyone loves a dancing goat, and yours can learn how.

The Dancer

It's fairly easy to teach your goat to dance, especially if he likes to stand on his hind legs. My Edmund has danced since he was a tiny bottle kid.

If your goat isn't inclined to stand on his own, begin by luring him with a favorite tidbit. When he stands, *without touching you with his feet,* click and reward. Don't reward him for standing with his feet against you; goat feet hurt and this can be a terrible habit to break. If your goat is one who naturally stands without luring, train him using free shaping: wait until he stands on his own, then click and reward.

Whichever method you use, begin rewarding for standing for longer durations of time and then introduce a verbal cue ("stand") or a hand signal (point, palm up, and sweep upward with your hand).

When he stands reliably for a reasonable length of time, teach him to turn. Lure him with a tidbit or target with your hand. Click and reward for small movements, as in teaching the spin (page 76), but don't expect great speed. Add a cue: say a word ("dance") or point, palm down, and rotate your hand in a circle. *Hint:* If you play a specific type of music as you begin each training session, your goat will associate it with the trick he's learning.

Congratulations, your goat can dance!

Wagons, Ho!

*The goat becomes the professor
whenever teachers can't be found.*

~ Turkish proverb

Why teach your goat to pull a cart? Why not? A trained harness goat can pull a decorated float in the kiddies day parade or help around the farm by hauling garden trimmings to the compost heap. You can drive your favorite goat to a cart on a balmy autumn day or harness a pair to give rides at a nursing home. Teaching goats to drive or pull a wagon or cart doesn't cost an arm and leg to get started, and it's easy: goats take to a harness like mice to cheese.

Working goat carts and wagons were a common sight a century ago. Studebaker made goat wagons, and early-twentieth-century dry goods catalogs such as Sears and J.C. Penney carried goat harnesses, wagons, and carts. Now they're staging a comeback, thanks to 4-H harness goat projects and organizations such as the American Harness Goat Association, the Canadian Pack and Harness Goat Association,

and British Harness Goat Society. Be part of the fun!

Harnessing the Right Goat

The goat you train for harness work should be sound of body and of mind. Both sexes can work in harness, though bucks can be awfully smelly during rut. The best harness goats are usually mature wethers. Whatever its sex, a harness goat should be well grown (at least a year old) before training begins, and calm and amenable to training.

All sizes of goats can pull an appropriately sized wagon, but if your goat is a little guy, his wagon shouldn't hold a passenger or a considerable load. Because carts are designed to carry one passenger (or two if they're children), only strong, full-size breeds should be used to pull a cart.

The best harness goats aren't sluggards, nor are they easily spooked. Most pet goats make fine driving goats if they're fitted with the right gear and then trained and conditioned properly to do the job.

Match the Load to the Goat

Goats can handle 15 to 20 percent of their own body weight on their back. How well the cart or wagon is balanced will determine how much weight is ultimately on an animal's back.

Try this: Ask a friend to sit in a typical cart, then pick up its shafts. If it's well balanced, there should be very little weight on the shafts when they're held at working height, parallel to the ground. Keep in mind that the point of balance changes as weight is added to the cart; good carts have a means of adjusting the balance point, so that even when loaded, little weight is put on the harnessed animal's back. A small goat hitched to a finely balanced cart can pull more weight than a bruiser hooked to a cart that places a great deal of pressure on his back.

Another factor to consider is the terrain that will be covered. It's easier to pull a vehicle on level terrain than up and down steep hills, and it's easier to pull on smooth surfaces than on rough, rocky pastures.

• • • • • • • • • • • • • •

He who lets the goat
be laid on his shoulders
is soon after forced
to carry the cow.

~ Italian proverb

• • • • • • • • • • • • • •

A well-balanced cart is a must. Check for proper balance before you buy one.

Still, there is a limit to how much weight a goat can manage, even when it is hitched to the best of carts. Don't strain your goat. If you want to pull a heavy load and you're sure that your goat can handle it, a four-wheeled vehicle is best. It removes most of the weight from the hitched animal's back and is easier to pull, and a plus: it's easy to convert an inexpensive utility wagon into a vehicle suitable for goats. Later in this chapter we'll show you how.

The Harness Explained

It's best to buy a harness specifically designed for goats. No amount of adjustment can make a driving bridle designed for a miniature horse fit a goat. If you do buy a standard miniature horse harness, however, be sure to remove the crupper (the loop that goes under an equine's tail).

Quality goat harnesses are made from leather and synthetic material, such as nylon, Biothane, and Beta.

DRESSED FOR SHOW

Top left and bottom right: *The ancestors of Britain's white military goat mascots were Kashmir goats from Queen Victoria's royal herd.*

Top right: *Goats have always been popular military and sports teams' mascots.*

Bottom left: *The author's Nubian wether, Uzzi, seems bemused by his Christmas finery.*

PACKGOATS

Packgoats are ideal for high-country camping. Climbing is what goats do best!

PULLING CARTS

Goats are used all over the world to transport people from one location to another. People race goat carts, kids enjoy short rides in them for fun, and adults can even use them to transport light loads.

BABIES

Clockwise from left to right: *Alpine, Boer, Alpine, Nubian*

AT PLAY

Top left and right: *The author's Nubian kids love to pounce on each other and climb on whatever is around the farm.*

Bottom three: *These Alpine goats and kids are very curious and interested in anyone who happens to be nearby.*

SHEARING

Fiber goats, except for Cashmeres and some Pygoras and
Nigoras, must be shorn, usually
twice a year. The task is best
accomplished with electric clippers but can be done by hand
using old-style hand shears or,
if you have only a few goats to
process and plenty of time, a
sharp pair of Fiskars scissors.

FINGER NURSING AND BOTTLE FEEDING

Bottle feeding can be a very rewarding experience. Amazingly, the Alpine kid below is nibbling the little girl's hand. Goat kids almost never suck on fingers!

CHEESE MAKING

Photos 1 and 2: *A goat with nice, hand-size teats like the ones on this Alpine is always a pleasure to milk. By using a partially covered milking pail, there is less chance of hair or other debris falling into the bucket.*

Photos 3 and 4: *Curds are being hung in cheese-cloth bags to drain.*

Photo 5: *Fresh goat cheese is pressed into individual containers for sale.*

A **leather harness** is beautiful and durable, but it's considerably more expensive than a synthetic harness and takes more effort to maintain. If you buy a leather harness, buy a good one. Quality leather feels smooth, supple, and substantial; avoid wrinkled, fibrous leather that stretches and breaks easily.

A **heavy-duty nylon harness** is strong, inexpensive, and easy to clean — stuff it in a pillowcase, drop it in the washer, and then hang it up to dry. The least expensive are usually made of nylon strapping.

Biothane is polyurethane-coated nylon, so it's strong and durable. It doesn't crack or peel, it won't absorb sweat, and it's easy to clean by simply hosing it off. It stays supple even in subzero temperatures. Biothane won't chafe your goat, and it molds to his shape with repeated use. It should outlast and outperform both leather and uncoated nylon, and it comes in a dazzling array of colors.

Beta is nylon strapping covered with a soft PVC vinyl coating instead of polyurethane. It comes in two thicknesses — $\frac{1}{16}$ inch and $\frac{1}{8}$ inch — and it looks and feels like leather. It's easy to clean (dip it in warm water and wipe it dry) and a bit more supple than regular Biothane. Beta has a softer coating, however, so it isn't as scratch resistant.

Harnesses should be sewn with small, neat stitches. When buying one used, make certain all stitching is sound. Good equipment is laced or stitched, not riveted. Choose a harness with solid brass or stainless-steel fittings; avoid weak "never-rust" (nickel-silver) fittings, and refuse a harness fitted with cheap, chrome-plated pot-metal hardware.

When properly adjusted, a harness should have extra holes for increasing and reducing size in the event your goat gains or loses weight or you'd like to use it on another goat.

Learn the Lingo

cart. A two-wheeled driving vehicle

carriage. A lightweight, four-wheeled driving vehicle

dash. The section of floor that curves up in front of the driver

header. Someone who jumps out of the vehicle, goes to the goat's head, and steadies the animal in an emergency

putting-to (also called hitching). The act of attaching a harnessed animal to a vehicle

shafts. Two metal or wooden poles attached to the front of a vehicle that are parallel to the goat's sides when he's driven

singletree (also called a whiffletree or swingletree). A wooden or metal piece with a swivel point in the center to which the harness traces are attached

wagon. A heavy-duty, four-wheeled vehicle (the pickup truck of the driving world)

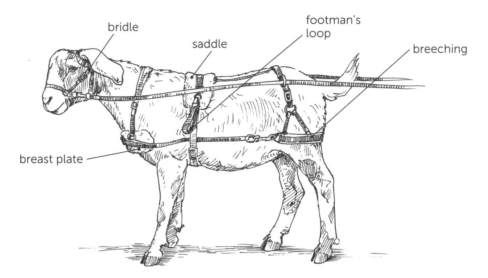

The Ins and Outs of the Goat Harness

At first glance, a harness has a mind-boggling number of parts. Each part has a job to do, so don't remove anything unless you know for certain you won't need it.

Headgear

A driving goat wears some sort of bridle or driving halter, which has reins that connect his head to the driver's hands. Goats that pull a wagon wear a standard halter or collar and a lead.

Well-trained goats respond to oral commands, but the bridle, halter, or collar is used for added control.

A bit-type driving bridle should be adjusted so that the bit rests lightly against the corners of the goat's mouth; the throatlatch should be snug enough to prevent the bridle from

Here are two common bits for goats: a bar mouth (top) and a snaffle mouth (bottom).

Goats can be driven in a bridle with a bit (left) or in a driving halter (right). Most goats greatly prefer the latter.

Is He Driving or Pulling?

A goat is driven by a person seated in a goat-drawn vehicle. He wears a full harness, including a driving bridle or a harness with long reins, as well as breeching (a harness that drapes across the goat's hips and fastens to the shafts) to act as the vehicle's brakes.

A goat pulls a vehicle or load wearing a collar or halter and a regular-length lead. He is directed by a person walking beside him or following on foot. A pulling harness may or may not include breeching.

It's easier to teach a goat to pull than to drive.

slipping off but loose enough for the goat to flex his neck; and the noseband (if it has one) should be tight enough to keep the bridle close to his face but loose enough to easily insert one finger laid flat between the lower jaw and the noseband. Some goat bridles are fitted with blinders (flaps that prevent the goat from seeing behind him, to keep him focused on the road ahead); goats don't need them, so you can chuck them. If you decide to use blinders, they should be centered over the goat's eyes and shouldn't brush his eyelashes. Most goat bridles have a solid mouthpiece bar bit or jointed mouthpiece snaffle bit with a 3-inch (7.5 cm) mouthpiece and 1-inch (2.5 cm) rings.

Most goats prefer a driving halter to a bridle and bit. Quality Llama Products (see Resources) makes a good one. The driving halter must be snug enough that it doesn't move around when cues are given but not so tight that it chafes the goat's head.

Reins

Reins connect the goat's mouth to the driver's hands. They fasten to the bit or side rings on his driving halter and pass through rings, called terrets, on the neck strap and/or saddle.

Breast Strap

The breast strap passes around the goat's chest just below the base of his neck. Adjusted too high, it presses against the windpipe and chokes the goat; too low, and much of the draft power is lost. The neck strap keeps the breast strap in the correct position, which is especially important because without it the breast strap would slip too far down the goat's slanted chest.

Traces (and the Singletree)

Traces, which are placed on either side of the goat, transmit pulling power from the breast strap to the cart or wagon. Traces are designed to be fastened to a ring or buckle on the breast strap or sewn directly in place. A slot in the opposite end of each trace slides over one end of the singletree on the cart or wagon. Usually a piece of leather holds it in place.

When the goat is hooked (harnessed and attached to a vehicle), traces should ideally lie in line with the vehicle's shafts. To adjust the height of the shafts, raise or lower the shaft loops on each side of the saddle; the traces' position is fixed by the breast collar and cannot be easily changed.

The singletree should be long enough to prevent the traces from coming in contact with the goat's body, and the traces should be long enough to prevent his hind legs from touching the singletree. When in motion, a goat's legs and shoulders move backward and

The vehicle you choose should be fitted with a barlike arrangement called a singletree. A movable singletree prevents the harness from causing painful sores on the goat's chest and shoulders.

forward and so do the ends of the breast strap, so the traces must be free to move as well. The singletree swings with the goat's shoulder movements, preventing the breast strap from rubbing sores on his chest and shoulders. Therefore, traces must not be fixed directly to a cart; this is why cheap carts without singletrees are bad news. The possible exception: a singletree isn't an absolute must when using a pulling harness attached to a wagon carrying a light load for a short distance — but even then, a wagon with a singletree works best.

Saddle and Girth

The saddle is designed to comfortably spread the vertical force of the shafts over the goat's back. The girth (also called the bellyband) prevents the cart from tipping back. The saddle and girth work together to balance the weight of the cart.

The saddle sits directly behind the goat's withers. The saddle and girth should be snug around the goat's belly but not tight. The center hook (also called the water hook), if it has one, is on top of the saddle. There are two rings on each side of the water hook. The reins go through these rings. On each side of the saddle are the shaft loops; these adjust to fit the shafts.

Breeching

Breeching acts as the vehicle's braking system. It prevents the vehicle from ramming into the goat's rear end, and it enables the goat to push the vehicle backward while backing up. Breeching fits around the goat's hindquarters and is held in place with hip straps. Breeching straps attach to the cart, going through the breeching dee, or footman's loop, a metal fitting on each shaft. The breeching strap passes through this metal fitting and wraps around the shaft.

The breeching band should lie halfway between the goat's tail head and his hocks. The hip strap prevents the breeching from falling down; it should lie just behind the point of the hip. When the goat is pulling, there should be roughly ¾ inch (2 cm) (depending on the goat's size) between the breeching band and his rump.

Whip

A wise driver carries a whip to help cue the goat's movements. It should be long enough to touch the goat's shoulders. It is never properly used behind the saddle.

Should You Make Your Own Harness?

It's possible to build your own driving and pulling harness if you use quality materials and you understand how a harness works and fits. If you don't, don't risk injuring your goat; buy a ready-made harness.

If you do choose to make your own harness, use a plan describing a harness for goats. *Do not use a plan for a dog harness!* It won't fit right and can seriously injure your goat. See Resources (page 199) for the names of some suppliers.

Cart with This

Most vehicles designed for donkeys and miniature horses can easily be converted for goats. Some dog carts adapt well, too.

However, you can purchase well-made, easy-entry metal carts specifically designed for goats. This is the type of cart I use. Since I clicker-train, I want to be able to get in and out of the cart with ease. Nonetheless, easy-entry carts do have their drawbacks. Their balloon tires are prone to punctures; their tubular metal shafts don't break but bend, which is a bad thing in a wreck; and some are poorly constructed using shoddy materials. No matter what type of vehicle you choose, keep these points in mind.

Easy-entry carts are designed so passengers step directly into one without having to climb over the longer shafts of a conventional cart.

Start with a vehicle of a recent make. Avoid the antique goat cart stored in Grandpa's barn or the cute but unrestored Studebaker goat wagon for sale at the Amish buggy maker's shop — at least until your goat is fully trained and you know how to evaluate and properly restore a vintage vehicle. Most old wooden vehicles demand complete restoration, and even newer ones usually need new (expensive!) wheels and shafts.

If you don't choose an easy-entry cart, make certain you and any passengers you plan to carry can enter and exit the vehicle safely and with relative ease. Make certain the cart you buy can be adjusted so that it's perfectly balanced with a passenger aboard.

Choose a vehicle with a singletree; cheap vehicles sometimes don't have them. Consider a vehicle with foot brakes. They aren't essential, but they're very, very nice to have, especially when traveling downhill.

Select a vehicle with a good suspension system. It equates with a comfortable ride. Easy-entry carts have coil springs; better wooden vehicles usually have leaf suspension. Given a choice, go for leaf springs — they absorb more shock.

Before you purchase a vehicle, determine whether its shafts will fit your goat. First measure your goat's length and his width at shoulder and hip, then measure his height at the withers. The proper shaft height/goat height ratio is 3 to 4. For example, the top edge of the shafts, held horizontally, should be 27 inches (68.5 cm) from the ground for a cart for a 36-inch (91 cm) goat. With carts, when you measure goat and shaft heights, it's best to have a driver on the cart and the goat in the harness, with the floor of the vehicle parallel to the ground.

Choose a vehicle with sturdy shafts. A vehicle with metal shafts should be fitted with welded end caps, not rubber stoppers that pop out and expose ragged ends.

If you'll be having various species pull your cart, purchase a vehicle with replaceable shafts. Then, by changing its tires to larger or smaller versions, you can use the vehicle with animals of various sizes. For instance, with a change of tires and shafts and a few other minor adjustments, one of my carts fits both my Boer goats and our miniature horses.

Converting a Wagon to Fit Goats

Because balance is such an issue with carts, it's best to buy a balanced cart instead of trying to design one. However, converting a utility wagon to fit goats is easy and fun.

Utility wagons are the kind of heavy-duty, relatively inexpensive metal wagons used for yard work. They're designed to be pulled by hand or by a riding lawn mower or tractor. My own cart is a 1,200-pound-capacity Gorilla Carts heavy-duty farm cart from TSC.

When choosing a wagon, look for center-pivot, automotive-type steering with tie-rods

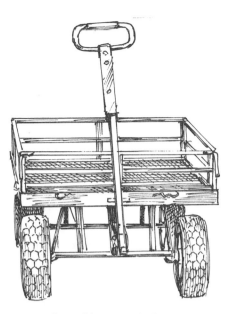

You can easily and inexpensively convert a utility wagon such as this one for goats.

(if you aren't sure what this is, ask a salesperson for help). Make certain the wagon has wheel bearings and pneumatic tires with tubes; tubeless tires tend to go flat, and once they do, it's hard to seal them back to the wheel. Look for a reasonably low-set wagon with a wide stance (the wheels should be set wide apart) to prevent it from tipping over. And, especially if someone will ride in the vehicle, select a wagon with a sturdy bed.

Converting a wagon is as simple as removing the handle (it's designed for easy removal) and attaching a shafts assembly. You can buy a shafts assembly that is ready-made or you can make your own (see my website, Dreamgoat Annie, listed in the Resources, for instructions).

Vehicle Care and Maintenance

Once you have a cart or wagon, make it last. Check the air in the tires before and after you drive, adding air as needed. Store your vehicle out of the weather — if you can't place it in a shelter, cover it with a fabric throw or canvas tarp. And at least twice per year (more often is better) do the following:

- Grease or oil the wheel bearings (if your wheels have enclosed bearings, you can skip this step).
- Wash the cart or wagon (including the wheels), dry thoroughly, and wax metal and wooden parts with the appropriate products.
- Treat the vinyl seat, if it has one, with vinyl conditioner.

Teaching Your Goat to Pull or Drive

Goats are followers — they expect to shadow humans, not the other way around. The hardest thing you'll do is teach your goat that you want him to walk in front of you, not behind.

This is my simple approach to teaching my animals, be they goats, donkeys, or mini horses, to pull or to drive. Here, clicker training shines. Click and reward him as he moves incrementally farther ahead. Remember: This is against his instincts, so give him plenty of time to understand what you're asking. The process may take anywhere from four or five short sessions to several months; the important thing is to make certain your goat understands each step before taking the next one. Goats are intelligent and very clever, so if you're patient and give them a chance to understand what you're asking, they learn easily.

Do the Groundwork First

To begin, grab your clicker and goody bag (if you use one), slip a halter or collar on your lead-trained goat, and take him someplace where he won't be distracted.

Start by shaping these behaviors: start, stop, stand still, reverse, trot from a walk, walk from a trot, and turn right and left. Then teach commands. You'll eventually need verbal cues: *get up, hup,* or *step out* to move out; *whoa* to stop;

Ground driving teaches your goat to walk in front of you; goats are naturally wired to follow.

stand or *stand up* to stand still; *back* or *back up* to reverse; *trot* or *trot on* to trot from a walk; *walk* to walk from a trot; and, traditionally, *gee* and *haw* to turn right and left.

Work every day until your goat fully understands each cue (remember: several short sessions are always better than one long one). *Whoa* is the most important command. Draw it out (*whoooooa*) so it sounds like no other word, and use it *only* when you expect him to stop. Don't use *whoa* once he's stopped and you want him to stand still (*easy* is a better word for that); *whoa* should mean one thing and one thing only: stop right now! Don't do anything else until he responds exactly as he should every time you say *whoa*.

Next, teach your goat to ground-drive. It's best to recruit a helper to walk beside your goat during the first few sessions to help him understand what you want him to do.

Show your goat the equipment he'll be wearing. Let him sniff it, then click and reward as you add each piece of harness until he's easy about wearing all of his gear. Remove everything but the halter or driving bridle and add reins. Click and reward until he's walking ahead of you with your helper at his side, starting, stopping, and turning with precision.

Enter: the Vehicle

Take your harnessed goat, wearing a halter with an attached lead, to an enclosed area and allow him to examine the waiting cart or wagon. Ask your helper to pull around the cart or wagon while the goat watches. When he doesn't mind that, ask your goat to stand still and have your assistant bring the vehicle behind him and lower the shafts. With you on one side and your helper on the other, walk him forward, *unhitched*, while the two of you pull the vehicle along in place.

Typically, a goat adjusts easily to this training process. If he doesn't, don't go any further until he does.

Keep these safety tips in mind when harnessing and working around harnessed animals of any species.

- If there is any doubt in your mind regarding how to harness and hitch correctly, ask an experienced person to show you how it is done.
- Keep one hand on a goat wearing blinders or talk constantly, so he knows where you are at all times.
- Never tie a harnessed animal by the bit.

Your goat learns not to fear the rattle of a cart behind him by walking between you and your assistant while you pull the cart.

- Don't leave your goat hooked to a vehicle unattended, not even for a minute.
- When hooking an animal, make sure he is wearing a sturdy halter over his bridle and is safely and securely tied to a sturdy object (not a tree limb, fence post, or car-door handle) using a sturdy lead rope.

Drive On!

The next step is to fit your goat with the appropriate headgear. For driving, this would be a driving halter with a lead attached or a halter and lead layered over his driving bridle. Never lead a goat with a lead attached to his bit because the bit will be pulled sideways and possibly injure his mouth. For pulling a wagon with the handler on foot, a halter or a collar and lead works just fine.

After he is fitted with headgear, hitch your goat to his vehicle and lead him around until he's comfortable with the vehicle following behind him.

Next, ask your assistant to walk near your goat's head as you drive him on foot. When he's comfortable with that step, either you or your helper can board the vehicle and drive him from that position. The other person should walk beside your goat's head. Repeat this procedure for several sessions, making certain your goat reliably stops, starts, and turns with ease. Then try driving him without your helper walking close to his side. Finally, take him out of the training enclosure. Congratulations, you've trained your goat to harness!

Now that you've safely mastered driving, you can do any number of things: join a driv-

Because goats prefer to follow rather than walk ahead of people, it's easier to train a cart goat when you have a helper to walk with him until he understands he's to lead while you follow, first on foot and then in the cart.

ing club or start a 4-H harness goat group; take your beautiful goat and his rig in parades; train another goat and have a team or train a bunch and create a four- or six-up team. Have fun with your harness goats!

A single serious or traumatic accident can scar your goat for life, so stay safe whenever you drive.

- Double-check your gear before you begin. Make sure that everything fits correctly and is in good condition.
- Quietly enter and exit the vehicle; don't dawdle, and never engage in horseplay.
- If carrying a passenger, especially a child, the driver (with reins in hand) should be the first one in the cart and the last one out.

- Never tie yourself or any passenger into a cart or wagon.
- Keep your hands off the wheels and make sure passengers do, too.
- Stay awake; don't let your mind wander. Watch your goat's body language. Always anticipate his actions and behave accordingly.

· · · · · · · · · · · · · ·

A goat with long horns, even if he does not butt, will be accused of butting.

~ Malay proverb

· · · · · · · · · · · · · ·

Nan

Nan, World War I mascot of the Canadian 21st Battalion CEF, poses with one of her friends, presumably Piper Nelson. Photo compliments of Al Lloyd and the 21st Battalion CEF discussion list (see Resources, page 198).

A VALIANT wartime mascot was Nan, caprine mascot of the Canadian 21st Battalion CEF. She served with the men of the 21st Battalion from its mustering in eastern Ontario in 1915 through its demobilization in 1919. During that time, 3,328 men of the 21st Battalion were killed, wounded, or went missing in action; only 106 of the unit's original soldiers — and Nan — entered Germany at the end of the war. Like Sergeant Bill (see "Sergeant Bill," page 128), Nan earned the 1914/1915 Star, the British War Medal, and the Victory Medal for her valor at the front.

When it came time for her battalion to move, Nan watched for the men of the Quartermaster's Detachment and Transport section to prepare her place on one of the general service wagons. Then, without prompting, she hopped aboard, ready for deployment.

During the march to the Somme, the unit's transport officer decided the men spent too much time caring for Nan and sold her to a French woman for 20 francs. When Nan's caretakers discovered she was gone, they were so horrified and outraged that they quickly found Nan, annulled the sale, and returned Nan to her place in the ranks. Nan was in many of the same battles as Sergeant Bill, and she was the first Allied mascot to cross the Rhine.

Despite her close contact with violence in the war, Nan's closest brush with death came at war's end, when her men unloaded her at Southampton, England. It was against Board of Agriculture regulations to bring into England an animal from a foreign country. Nan, the board insisted, would have to be slaughtered or deported. After three weeks in quarantine, she boarded the Cunard liner *Carolina* with the men of the 21st Battalion and departed for Canada.

After her battalion was disbanded, Nan spent the rest of the summer on the grounds of Mowat Hospital, where she was cared for by Piper Nelson. In the fall she moved to the Royal Military College in Kingston, Ontario, where she lived in the stables for the rest of her days. On September 22, 1924, at the ripe old age of 12, Nan lost the use of her legs; she was painlessly put down and buried by veterans of her battalion.

A corner of the military museum at the Armouries in Kingston, Ontario, is devoted to Nan's military career. Read all about Nan and her participation in the war effort at the Canadian 21st Battalion CEF website (see Resources, page 198).

CHAPTER

Packing with Goats

Even the humblest goat is useful.

~ Haitian proverb

Goat packing — hiking with goats who carry the packs — is a rapidly growing recreational pastime in the mountains of the West, where proponents breed strong, long-legged goats specifically for the task. However, any willing goat makes a great packgoat for casual packing.

And yes, the correct term is *packgoat,* not *pack goat,* according to the North American Packgoat Association (see Resources), a group of gung-ho packgoat enthusiasts formed in March 2001 to promote goat packing and to keep goat packers abreast of national forest policies pertaining to packing with goats. The group also hosts Goatstock, an annual four-day rendezvous of packgoats and their people in one of the western states, as well as a sanctioned rendezvous farther east. The North American Packgoat Association website is a rich source of information about everything packgoat — don't miss it!

Why Pack with Goats?

There are lots of good reasons for hiking and camping with pack animals. Older folks whose bodies are no longer up to toting heavy backpacks (that's me) can continue camping in backwoods places. Families with babies and toddlers can camp there, too: goats carry the gear while Mom and Dad pack the youngsters on their backs. And anyone can benefit from an extra packer that can haul some amenities: a large goat can tote a good-size load of gear, including a lightweight tent, folding camp cot, and sleeping bag, with ease.

Llamas and donkeys (usually miniature donkeys) can also be used as pack animals, but there are some downsides to their use. Donkeys' larger, one-piece hooves and bigger droppings aren't compatible with low-impact travel; they aren't adept climbers; and a donkey must be led along the trail. Llamas' padded toes and elklike droppings are environmentally friendly, but their aloof, catlike ways don't appeal to some people; and llamas have to be led as well.

Goats are friendly, happy packers. They're inexpensive to buy and to keep, their gear is less expensive than llama and donkey gear, and they travel off-lead over rough terrain. A goat can go anywhere a human can (and then some) — over packed snow and downed logs, and up rocks and steep inclines. Goats are strong; a well-conditioned adult goat can pack 25 percent of his own weight, covering 5 to 15 mountainous miles in a day. They are short enough that even children can saddle goats and load their packs, and you can transport goats to the trailhead in a trailer, a carrier in the bed of a truck, or even a van or SUV.

Goats' impact on the environment is virtually nil. Their hoofprints and droppings are much like those of deer, and they distribute their poop as they walk, unlike llamas, which use only large, communal dung piles. Goats provide most of their nutritional needs by browsing — a bite here, a nibble there — as they move along the trail. Because they're environmentally friendly, goats are welcome in most national forests and on Bureau of Land Management lands, and many state parks allow packgoats on their horse trails.

Which Goat Is a Good Packgoat?

Any goat can pack if he's built for the job at hand and mentally fit. He or she should be healthy and physically fit, with a broad chest, well-sprung ribs, heavy bone, strong hooves, and a fairly level back, and have a calm temperament. Both wethers and open (nonlactating) does make fine packgoats; bucks are fine for part of the year but stinky and obnoxious during rut. Most people pack with wethers. Most goat packers prefer horned goats, but this is really only a personal choice.

Age is important; goats shouldn't carry significant weight (up to 25 percent of their body weight) until they're at least 3 years old. Kids and yearlings often go packing with their peers but without carrying a pack; they hike alongside them, learning trail manners in the process. Some folks use lightly loaded soft packs on yearlings and fit 2-year-olds with regular packing gear that holds 10 to 15 percent of the goat's weight. Since well-cared-for, properly conditioned packgoats can work until about 12 years of age, it pays to go easy on them as youngsters.

Just as important as age is proper attitude. Only a goat who has bonded to humans will follow off-lead. Bottle babies and highly socialized dam-raised kids invariably make the best packgoats. And the goat must *want* to pack. Unhappy packers tend to lie down on the trail. Packers call great attitude "gung-ho goat."

Seasoned western goat packers prefer dairy-breed wethers, with the exception of Nubians, which they say are too vocal on the trail and have a poor work ethic. Others love the company of Nubians. My Nubians are noisy, true, but they're hard workers; it all boils down to personal preference. Western packers' preferred breeds appear to be Saanens, LaManchas, Oberhaslis, Alpines, and Toggenburgs, in that order. Most favor crosses and some like to have a hint of Boer mixed in.

The trick is to pick goats to suit your needs and terrain. I like Boers for their ability to

A training pack is best for short trips carrying very light loads, such as heading out for a picnic in the woods.

A sawbuck saddle is perfectly engineered for longer trips.

carry heavier loads and for their easygoing nature. Their longer bodies and short legs work fine where I hike in the Ozark Mountains of Arkansas and Missouri, but they wouldn't work so well in rugged western goat packing. The other meat breeds, and even fiber goats such as shorn Angoras, make great packgoats for packing in level to heavily rolling terrain.

Outfitting Your Packgoat

Your goat will need a collar or halter (see chapter 9), a lead, and a pack. Pack styles include training packs; companion packs; and full-fledged saddle, pad, and pannier combinations. Good packing gear of any type provides adequate padding, alleviates pressure on the spine, and is very stable, allowing for slightly different weights in each saddlebag without shifting to the side or moving while the goat walks.

A training pack is a simple, soft-sided, saddlebag-like affair with a single girth and built-in chest and butt straps to hold it in place. The fabric connecting the two bags lies directly on the goat's back, and this is not a good thing — too much weight pressing directly on any pack animal's spine over a prolonged period of time causes permanent damage. Training packs are best used for carrying light loads on day hikes, nothing more (many veteran goat packers recommend never using them).

A companion pack is a better choice for longer day trips and overnight camping. A good companion pack features a thick, divided pad that keeps the weight off its wearer's spine, detachable panniers like those used in full-fledged packing gear, a single girth, and a built-in chest strap and a butt strap to keep it in place. These outfits, complete with pads and panniers, cost more than training packs and are harder to find.

Packsaddle outfits include two detachable panniers and a packsaddle secured with one or two wide girths in the middle, a chest strap in front, and a butt strap behind. The saddle provides greater stability and complete spinal relief, making it possible for a goat to carry considerably heavier loads.

A Closer Look at Packsaddles

Packsaddles come in a variety of materials and configurations, from do-it-yourself, wooden sawbuck kits to newfangled designs made with aluminum instead of traditional wooden bows. Most weigh 4 to 8 pounds (1.8–3.6 kg) with the pad.

They come with leather, nylon web, vinyl-covered nylon web, or Biothane strapping.

A companion pack falls somewhere between a day pack and a sawbuck-style packsaddle. It's great for day trips and overnight camping.

Saddling Your Goat

Saddle your goat correctly to prevent saddle sores and strapping rubs.

1. Groom your goat, paying special attention to his back, chest, and butt.

2. Place the saddle pad 6 inches ahead of its proper position and slide it back into place (this smooths the hair on your goat's back).

3. Center the saddle on the pad. It should never sit on a goat's shoulder blades or hit his spine at its back edge.

4. Fasten the girth. It should be snug but not crushing. If you can slip two fingers between the goat and the girth strap, it's just right.

5. Buckle the chest and butt straps. Never fasten these before fastening the girth; if he moves away, the saddle could turn and be damaged by your frightened goat.

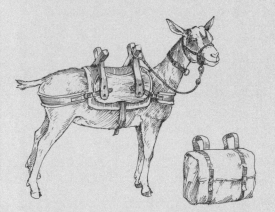

Pannier loops fit over the bows of a sawbuck packsaddle, making saddling up a snap.

The Three Billy-Goats Gruff

This is a famous Norwegian folktale ("De tre bukkene Bruse"), one of many collected by Peter Christian Asbjørnsen and Jørgan Moe and published in the mid-1800s. This version is one of the earliest in print.

• • •

ONCE UPON A TIME there were three billy-goats, who were to go up to the hill-side to make themselves fat, and the name of all the three was Gruff.

On the way up was a bridge over a burn they had to cross; and under the bridge lived a great ugly Troll with eyes as big as saucers, and a nose as long as a poker.

So first of all came the youngest billy-goat Gruff to cross the bridge.

"Trip, trap; trip, trap!" went the bridge.

"Who's that tripping over my bridge?" said the Troll.

"Oh! It is only I, the tiniest billy-goat Gruff; and I'm going up to the hill-side to make myself fat," said the billy-goat, with such a small voice.

"Now, I'm coming to gobble you up," said the Troll.

"Oh no! pray don't take me. I'm too little, that I am," said the billy-goat; "wait a bit until the second billy-goat Gruff comes, he's much bigger."

"Well! Be off with you," said the Troll.

A little while after came the second billy-goat Gruff to cross the bridge.

"TRIP, TRAP! TRIP, TRAP! TRIP, TRAP!" went the bridge.

"WHO'S THAT tripping over my bridge?" roared the Troll.

"Oh! It's the second billy-goat Gruff; and I'm going up to the hill-side to make myself

fat," said the billy-goat, who hadn't such a small voice.

"Now, I'm coming to gobble you up," said the Troll.

"Oh no! don't take me. Wait a little until the big billy-goat Gruff comes, he's much bigger."

"Very well! Be off with you," said the Troll.

But just then came the big billy-goat Gruff.

"TRIP, TRAP! TRIP, TRAP! TRIP, TRAP!" went the bridge, for the billy-goat was so big and heavy that the bridge creaked and groaned under him.

"WHO'S THAT tramping over my bridge?" roared the Troll.

"It's I! the big billy-goat Gruff," said the billy-goat who had an ugly, hoarse voice of his own.

"Now, I'm coming to gobble you up," said the Troll.

"Well, come along! I've got two spears;
And I'll poke your eyeballs out at your ears;
I've got besides two curling stones,
And I'll crush you to bits, body and bones."

That was what the big billy-goat said; and so he flew at the Troll and poked his eyes out with his horns, and crushed him to bits, body and bones, and tossed him out into the burn, and after that he went up to the hill-side. There the billy-goats got so fat they were scarce able to walk home again; and if the fat hasn't fallen off them, why they're still fat; and so —

"Snip, snap, snout,
This tale's told out."

— A selection from *The Norse Tales for the Use of Children, by* George Webb Dasent

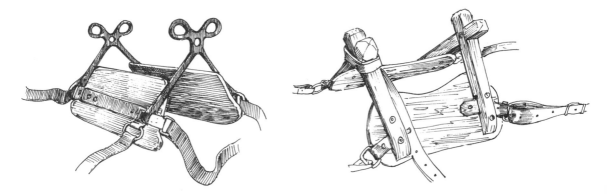

Packsaddles range from modern adaptations (left) to traditional sawbuck models made of wood (right).

Leather is beautiful and very traditional but needs more upkeep than synthetic strapping. Some synthetic strapping is lined with felt; this is comfy for the goat but harder to keep debris-free on the trail.

Traditional sawbuck packsaddles (also called cross buck packsaddles) resemble the ones used on pack mules. Bows (X-shaped front and back pieces) are usually crafted of oak or poplar. Sideboards (the flat pieces in contact with the goat) are made of strong but lightweight poplar, pine, or spruce and are set at roughly an 86-degree angle. Nontraditional packsaddles can be adjustable for a custom fit and made of aluminum.

No matter the materials used, sideboards should be beveled to allow for clearance at the goat's shoulder blades and hip bones and to help prevent them from rubbing the goat.

Pad Your Packsaddle!

A loaded packsaddle must always be used with a pad so it doesn't dig into its bearer's back. Some come with a pad; others don't. Some have pads with pockets that secure sideboards in place; a few have pads that are permanently attached.

Pads must be thick enough to do some good. Poke your pad and see if you can easily feel your finger on the opposite side. If you can, buy or make a thicker pad or use a second pad under the first one.

Pads can be cut from horse saddle pads, giving users a wide choice of materials and sizes. Wool felt horse saddle pads are hard to beat, and you can cut two packsaddle pads from each one.

Select a Soft Lead

Lead ropes for the trail should be soft, about ½ inch (1.5 cm) in diameter, and 5 to 7 feet (1.5–2 m) long. These are easy on your hands and won't injure your goat's legs should he somehow get them tangled in the lead.

When hiking with your goat off-lead, leave the rope attached to his halter or collar and fasten the free end to the saddle; that way you can restrain your goat quickly if necessary.

Pick Proper Panniers

Panniers, also called pack bags, carry your gear. They are secured to the packsaddle's bows by loops, so they're ultra-easy to use. Panniers come in an array of materials and sizes, so it's easy to find fabrics and colors you like. Waterproofed Cordura is my choice for durability and style.

Good panniers must be built to withstand life on the trail. They are sewn together with small, tight stitches, and webbing should be strong and securely sewn in place. Hardware such as buckles should be made of quality materials and reasonably easy to work with one hand. Outside pockets for maps, water bottles, and other often-accessed necessities are quite useful.

Some people use a bucket pannier. It uses straps to secure a bucket to the packsaddle. These are great for carrying easily damaged items such as cameras and food (especially eggs!).

Just as important as choosing the right pannier is packing it properly. This is both for your goat's safety and comfort and to protect the items in the pannier. When you pack for your trip, remember that heavy items go near the bottom of the pack. A top-heavy pack is unstable. It's also a good idea to place soft items on the side nearest your goat to act as additional padding, and, conversely, to pad pointy objects and pack them on the side farthest away from the goat.

Each pannier *must* weigh within a pound or so of the pannier on the opposite side of the saddle; otherwise you'll stress your goat's back and likely cause hard-to-heal saddle sores. Weigh your panniers as you load them and buy a small hand scale to pack along on overnight trips.

And last, when loading up, support the weight of the first pannier until the second one is on; otherwise you're likely to twist the

Nannyberries

GOATS AND THE TYLWYTH TEG

Among the traditions of the origin of the Gwyllion is one which associates them with goats. Goats are in Wales held in peculiar esteem for their supposed occult intellectual powers. They are believed to be on very good terms with the Tylwyth Teg (Welsh fairies) and possessed of more knowledge than their appearance indicates. It is one of the peculiarities of the Tylwyth Teg that every Friday night they comb the goats' beards to make them decent for Sunday.

— *British Goblins: Welsh Folk-lore, Fairy Mythology, Legends and Traditions,* Wirt Sikes

saddle. If you do, unsaddle and start over rather than pushing the saddle back in place.

Try Top-Stuff Sacks

Top-stuff sacks are bags used to load gear atop the bows of your saddle. They should never carry more than one-third of your load; otherwise, the goat's saddle will roll from side to side as he walks. A rolled-up sleeping bag, a folding camp cot, or any other piece of compact, lightweight gear can also be secured between the bows in lieu of a stuff sack.

Primer on Packgoat Training

Training goats to pack is easy; they're naturals. If your goat is bonded to you and trained to calmly lead and tie, it's simply a matter of introducing the equipment, practicing a bit at home, and then hitting the trail. But first, dig out your goody bag and clicker.

Saddle Up

Halter or collar your goat, attach a lead, and take him to where his saddle and pad are waiting. Pick them up one at a time and let him take a closer look. Let him touch the items with his nose and then click and reward. If he wants to nibble, that's okay, too.

Next, quietly place the pad on his back and click and reward. If he's concerned and steps away before the pad is on his back, gradually bring it closer, clicking and rewarding, until he's okay with it being on his back.

Then place the saddle on his back. Secure the cinch snugly but not too tight. Most goats take saddling in stride, but if he's going to object, it will be now. Fasten the chest and butt

Hiking with Four-Legged Kids

If your kid is too young to pack, take him hiking anyway. Teach him to lead, tie, and "stay" when told. He can tag along with your fully packed goat or just keep you company, and he'll learn trail manners as he goes.

straps, again clicking and rewarding for good behavior. Lead your saddled goat.

After your goat takes those steps calmly in stride, add panniers at the next session. Stuff them with paper the first time. Proceed as you did when saddling, clicking and rewarding until he's fully loaded.

Now lead your fully packed goat over varied terrain, taking care that he hears things brushing his panniers as he passes by. This sound frightens some goats, so be prepared. Eventually he will become accustomed to the noise.

When he's easy about wearing his equipment, begin adding weight to the panniers. Try beginning with 5 pounds (2¼ kg) and adding 2 more pounds (1 kg) each time until he's carrying close to what his maximum load would be. Don't overdo it; he'll need to become conditioned through working on the trail before he can carry the maximum load for his age.

Finally, unsnap the lead and allow him to follow untethered, clicking and rewarding him as he does. Practice at home a few more times, loose and fully loaded, before taking him packing.

Teach a goat to "stay" by holding out your hand as you take a step back.

Stop and Stay

Before you hit the trail, teach your goat to stand still when you ask him to. Walk with him at your side, then stop, step in front of him, and hold out your hand, palm forward, and take a big step back. If he stands still for even a heartbeat (if you hold your hand in this manner, most goats will), click and reward and then step back to his side and start walking again. Practice until he stands for at least several minutes, and then add your cue word (*stand* or *stay*).

Crossing Water

Goats naturally hate to get their feet wet, so teach your goat to walk through water at home. Put on your boots, then find or make a nice puddle. Snap a lead on your goat's collar and bring him closer to take a look. Lead him purposely toward the puddle; at some point he'll stop, plant his feet, and look at you like he thinks you're crazy.

Urge him forward (sometimes asking him to target will do the trick). At the first sign of movement, click and reward. Continue urging movement, then clicking and rewarding until he's fully in the puddle. This may take one session or a dozen, but when it does happen, reward him with a jackpot! Then click and reward for being led calmly out the other side. Later, repeat these steps on the trail, substituting real water for a puddle. It works!

Load Up

In most cases you'll haul your goat to the trail-head. If you have a horse trailer, training him to load is easy: just lead him in (gradually if he's fearful), clicking and rewarding for good behavior. Remember, however, that goats tend to stress when hauling solo, so take along a second goat if you possibly can.

If he'll travel in the back of a truck in a goat tote or truck topper, or in a van or SUV, teach him to jump in without a fuss. Start by luring your goat with food or by molding (placing his front feet where they need to go). Click and reward his gradual progress, and soon he'll be leaping in without a qualm.

Keep in mind, however, that some goats have physical limitations. Some short-legged, muscular goats, such as Boers and Savannas, can't jump terribly high. If you have such a goat, build a sturdy pedestal to act as a booster step, and make it secure so it doesn't tip over.

Trail Etiquette

In most places it's okay to hike with your goat off-lead. However, know and obey the rules before hiking in a state park or a state or national forest.

You'll probably be sharing trails with humans, horses, and other pack animals, so keep your goat near you at all times. He'll fall behind to browse but rush to catch up with you again; this is okay, just check frequently to make sure he's following (goats have been known to follow hikers headed in the opposite direction).

Teach your goat to walk behind, not in front of you. Carry a walking stick and if he tries to barge past, use it to block his way. When he falls back in line, click and reward him. He'll soon understand where you want him to be.

Trail etiquette indicates that pack animals have the right of way before hikers, bicycles, and ATVs. Ridden horses, donkeys, and mules have the right of way before pack donkeys, llamas, and goats. Grab your goat's lead and move to the side when meeting equines coming the other way; many horses spook at the sight of a goat.

Your goat should stay near you in camp. In group camp situations, don't let your goat visit neighboring camps mooching for food, even if they encourage it. If he's invited to visit their camps, take him on a lead. Don't relax your rules from home by allowing him to gorge on unearned treats, even when offered from other people's hands. Show them how to ask him to do a simple action to earn a treat. They'll be impressed with his intelligence, and your goat won't pick up bad manners.

Be sure to store your panniers where he can't help himself to any goodies you've stowed inside. Goats have also been known to munch

Dogs on the Trail

A persistent dog can kill or maim a goat with lightning speed, so be prepared to protect your goat from loose dogs on the trail. A rap with your walking stick might not be enough. Northwest Pack Goats sells a battery-operated, hand-held unit called a Dog Dazer that repels dogs using ultrasonic sound. A better choice where it's legal: pepper spray. Carrying proper deterrent could save your goat's life.

or mangle camping permits, trail maps, and paperback books. Make your panniers strictly off-limits.

Many goat packers allow their goats to forage off-lead at night, but it's a very bad idea. Predators abound in the backcountry and a free-ranging goat is at risk. Let him roam nearby while you're awake, but tie him when you go to bed. Use a rope that is long enough to enable him to lie down in comfort but not long enough to enable him to tangle his rope with the rope of another animal.

There may be something in life more fun than packing with goats, but I haven't found it yet. Get your goat and give it a try. I think you'll become hooked, too!

CHAPTER 7

Got Milk?

If you want milk, feed the goat.

~ Jamaican proverb

Why milk a goat? For me, the joy of spending quiet time twice a day in the presence of happy goats heads the list, but there are lots of other benefits, too.

Goat's milk contains less fat than the other types of milk consumed worldwide; it has three times more health-giving, medium-chain fatty acids than cow's milk; and it has more calcium, protein, riboflavin, and vitamins but less lactose and saturated fat.

Goat's milk is also more easily digested than cow's milk due to low levels of alpha s1-casein, a protein involved in curd formation. The smaller milk fat globules in goat's milk contribute to increased digestibility.

And goat's-milk cheese, besides being very creamy and delicious, can also be a profitable niche for people who want to make some money from their goats. There is a growing demand for artisanal cheese across the country.

Milking isn't for everyone, however, so make sure you want to do it before you commit. You must milk twice a day at the same time of day, every day, with no respite. If your doe doesn't produce much milk or she's nursing kids, you can milk once a day, but you'll have less return. Finding a farm sitter who is willing and able to milk is a formidable task; if you milk, you had best be a homebody or willing to schedule day trips and vacations around milking times and drying-off periods. Milking your goat will involve commitment, but the rewards are oh, so delicious!

Choosing Your Milk Goat and Supplies

We talked about conformation factors to consider when choosing goats in chapter 2, but this is important, too: choose a goat that you *like*. You'll spend a fair amount of time every day, day after day, in the company of the doe you milk; if she's pretty, clever, or appealing to you in any other way, you'll appreciate that time all the more. Neither of my Nubian dairy queens would rate high on a commercial milk line or in the ring at a dairy goat show (one of Bon Bon's teats is serviceable but twice the length of the other!), but I love them still. And, because you can't have milk if your doe doesn't have kids, I'll discuss kidding and taking care of young kids in part 3.

The Right Milk Goat for You

If your needs aren't great, non–dairy goats make great milk goats, too. Boer does give respectable amounts of high-butterfat, luscious milk, and people milk Angoras, Spanish does, and even Pygmy goats (they give rich, tasty milk). So don't discount a doe just because she isn't the "right" breed, she's lacking in conventional dairy character, or she doesn't sport a show-stopping udder. If you like her, you *like* her. That counts for a lot.

When shopping for a dairy goat, ask if you can milk her before you buy. If you don't know how to milk, ask the seller to milk her while you observe. Here are some pointers of what to look for.

- Make certain she is well behaved on the milking stand or at least doesn't have habits you can't live with.
- Grasp her teats. Will they be easy to milk? Tiny teats are difficult to manipulate; huge, sausage teats will tire your hands.
- How fast does the milk stream out? Goats with extra-large orifices are easy to milk, but because those openings allow

Nannyberries

SHEEP OR GOAT?

Tails It Is
Unless it's frightened or ill, a goat's tail sticks up; it's short, with a cute fringe of longer hair at the sides. Sheep's tails always hang down. Most wool sheep are born with a long, woolly tail that is docked (shortened) to help prevent flystrike later on.

What About a Beard?
Goats of both sexes can have beards, although not every goat has one. Sheep never have beards, although rams of most hair breeds have manes: longer hair on their shoulders and their lower neck.

Check the Lip
A sheep's upper lip is divided by a distinct, deep groove. There is a shallow crease in a goat's upper lip, but it's only superficial.

bacteria and staph organisms to enter easily, they're prone to developing mastitis. Yet, it takes forever to milk goats with tiny orifices. Look for a balance, and be aware that there should be only one opening per teat.

- And don't forget to taste her milk. Don't end up with a doe that gives strange-flavored milk!

A Milker's Supply List

If you're sure you want to milk, you'll need the right equipment. Milking machines for goats are available but pricey and slower than hand milking, given that the equipment must be sanitized after each use. The jury is out on handheld, suction-type milkers. They're okay for occasional use but probably damage udders when used on an ongoing, twice-daily basis; it's best to milk by hand.

You'll probably want a sturdy milking stand. It's a great back saver and allows you to see what you're doing. But don't despair if your goat won't climb up on one. Latifah is afraid of the stand (apparently having fallen off a rickety one sometime in the past), so I sit cross-legged on the ground to milk her. It works for us!

Nontoxic Fly Repellent for the Milking Stand

It's easy to make a safe, effective fly repellent to spray directly on your goats — and it costs just pennies!

1 cup distilled white vinegar
1 cup water
⅓ cup blue Dawn dish soap

Pour the ingredients into your sprayer, shake, and spray! Expect some soap buildup on your goat's coat. To remove it, give her a weekly bath (no additional soap is needed, just rinse her with a bucket or hose).

Here are some other things you will need:

- Stainless-steel pail or bowl to catch the milk (plastic receptacles can't be properly sanitized and glass can shatter if someone steps on it)
- Strip cup or dark-colored bowl
- Pre-milking teat cleanser or unscented baby wipes
- Post-milking teat dip or an aerosol product such as Fight Bac
- Strainer sized to hold 6½-inch milk filters
- Funnel
- Containers to hold milk in the refrigerator
- Plenty of soap and hot water to keep everything squeaky clean (cleanliness is the key to producing stupendous-tasting milk)

Let's Milk a Goat!

Most small-scale home milk goat owners allow kids to nurse freely for their first two weeks of life. After that, they are placed in their own separate quarters in the barn at night, where they can see their mom but not partake of her milk. If you choose this route (we do), feed your doe in the evening but don't milk her. At the same time, give the kids their own little bit of hay and a smidgen of grain. Being smart little goats, they learn the routine in days and soon scamper to their quarters with delight. Milk your doe in the morning and let her kids rejoin her for the day. Kids are normally weaned around 12 to 16 weeks of age. After that, the milk is all yours.

In an ideal world, you'd milk your goat at 12-hour intervals. That isn't always possible,

MILKING PROTOCOL

1. Swab your goat's udder with a paper towel dipped in teat cleanser and wrung nearly dry, or clean her teats with a baby wipe — the trick is to thoroughly clean them without saturating them with fluid.

2. Massage the udder for 30 seconds to facilitate milk letdown.

3. Aim several squirts of milk from each teat into the strip cup or bowl, one teat at a time, to test for stringy substances, blood, or anything out of the norm. If you find something strange, don't use the milk until you've tested for mastitis (see page 183). Most bacteria are contained in the first few squirts, so these should be discarded in any case.

4. Place your milking pail under the udder. Wrap your hand around a teat. If the teat is a large one or fat, hold it near its end. If it's short, don't include the udder itself in your grip (this hurts the goat and ultimately damages her udder); simply wrap as many fingers as you can around the teat, leaving your pinky and any additional fingers suspended in the air.

5. Gently nudge the teat upward with the side of your hand (this fills the teat reservoir). Tighten your thumb and forefinger to prevent milk from flowing back up into the udder, then successively tighten your middle finger, ring finger, and pinky, propelling the milk out into your pail. *Never* pull the teat and *never* strip it between your thumb and forefinger as cow milkers are wont to do (both practices are painful and damage a goat's more delicate teats).

6. Milk until both teats are flaccid, then stop and gently massage the udder for 30 seconds to facilitate final milk letdown. Alternately, nudge the udder with the sides of your hands as you milk those last few squirts (you're simulating kids bunting their mama's udder). Continue until you're coaxing out only tiny squirts of milk, if any.

7. When you're finished milking, dunk the ends of each of your doe's teats in teat dip or spritz them with Fight Bac, making sure the orifice is coated.

8. Allow the doe to hop down. Post-milking teat treatments such as Fight Bac tighten the orifice, but it will still be another half hour or so before it has fully rebounded. Since mastitis-causing bacteria are able to enter a slack orifice, try to prevent your doe from lying down in the interim. Offer her some yummy hay to keep her occupied or turn her out with her friends to graze.

9. Grab your pail and head to the house; or, if you're milking more does, dump the milk in a communal stainless-steel container, cover it (a hand towel works nicely), and proceed to the next goat.

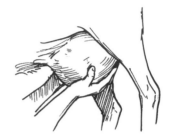

1. Secure the goat in the milking stand with a pail positioned under the udder.

2. After cleaning the udder, massage it.

3. Be sure your hand position on the teat is correct.

4. After milking, dip each teat in disinfecting teat dip.

but do allow as much time between morning and evening milkings as you can. The most important thing is that you milk at the same time each day.

It's best to milk in an area separate from your goat's living quarters. A separate milking area is easier to keep tidy and that way bedding, airborne dust, and vexatious flies are less likely to find their way to the milk pail.

At the appointed hour, scrub your hands, gather your pail and strip cup or dark-colored bowl, and head to the barn. Measure your doe's grain ration and deposit it in the grain cup on your milking stand, then lead her to the milking area. When she hops onto the stand, fasten her neck in the stanchion. Quickly brush her with a soft brush to dislodge loose hair, dirt, and bedding — you don't want to find these in the milk pail. Arrange your milking stool in a comfortable position on either side of your goat (most people milk on the goat's right side, but do what suits you best).

When milking, place the milking pail slightly forward of your doe's udder, then keep an eye on her hind legs! An innocent stomp to dislodge an annoying fly will tip the bucket as thoroughly as a well-aimed kick. While you're learning, it's wise to milk into a small receptacle you can dump into a pail from time to time.

Milking seems awfully hard at first and your back, fingers, and wrists may ache (a lot). Don't worry; coordination, speed, and strength come with practice. In a week or two you'll be working alternate teats in a rhythmic, easy pace you could barely fathom when you first began.

Processing and Storing Tasty Milk

Fresh, properly handled milk from healthy goats is naturally creamy, sweet, and tasty. To ensure that you produce the best-tasting milk you can, process the milk as hygienically and quickly as possible. Immediately after milking, strain it through a milk filter into glass containers (if storing it in the refrigerator) or plastic (if freezing it). For best results, submerge the containers in a sink of icy water for half an hour before refrigerating or freezing. This will prevent the milk from developing any off-flavors.

After each milking, be sure to sanitize your pails and milking paraphernalia in hot, sudsy water. Rinse thoroughly and upend them in a drain rack to dry.

A noisy goat gives little milk.
~ Polish proverb

Goat Milk, Goat Milk, Raw! Raw! Raw!

Must you pasteurize the milk you drink? That depends. Some authorities claim you'll be dead in weeks unless you do; others tout raw milk as a healing ambrosia of the gods. The truth lies somewhere in between.

My goats are healthy and I cool and bottle their milk right away (without pasteurization). It's perfectly okay, however, if you decide to pasteurize your milk. If so, you can buy a home milk pasteurizer to do the job or pasteurize small batches on the kitchen range.

Keep the Milk Flowing

Most does must be rebred each fall to continue producing milk; the few who don't are said to "milk through" (this depends on breed and bloodlines; it's fairly common for Swiss breeds to milk through). Does require at least 2 months of downtime to recuperate before giving birth and starting up milk production again, so if you need goat's milk year-round, you'll have to freeze some milk in the month(s) preceding the drying-off time or keep at least two does and stagger their breeding times.

Flash-pasteurization is an easy way to pasteurize your milk on the stove, but the resulting milk can be kept in the refrigerator for only a few days before it begins tasting "goaty."

To flash-pasteurize, place several inches of icy water in the sink. Fill the bottom of a double boiler with water and pour milk into the inner pan. Heat the milk to 161°F (71.7°C) (use a thermometer) and hold that temperature for 30 seconds, stirring constantly to prevent scorching. Then, set the inner pan in the sink full of icy water and allow the milk to cool. You may need to add more ice if your milk is still warm once all of the ice has melted.

When Pink Doesn't Mean Strawberry-Flavored

Blood clots in the milk? Don't automatically think the worst. First, run a CMT test to rule out mastitis (see page 183), then try to figure out what else might be wrong. First fresheners sometimes produce pink-tinged milk as their udders adapt to milking. Pink milk can also indicate a metabolic disturbance caused by lack of available blood calcium. A blow to the udder can do it, too.

It's fine to feed pink, nonmastitic milk to bottle kids or to any other type of animal you may be feeding. It's safe for humans, too, though it has a slightly "off," somewhat coppery taste. To partially clear up pink milk, allow it to sit in the refrigerator overnight or until blood spots settle to the bottom, then pour off the top level and use it however you like.

You can also slow-pasteurize milk by gradually heating it to 145°F (62.8°C) and holding it at that temperature for 30 minutes before cooling it quickly. Slow-pasteurized milk should stay fresh in the refrigerator for about 1 week, the same as for fresh, raw milk.

You Drink What She Eats

The forage and feed a cow, doe, or ewe eats and the scents she inhales can flavor the milk she gives you. According to the University of California Cooperative Extension's bulletin "Milk Quality and Flavor," 80 percent of the off-flavors in goat's milk are feed related (among other factors are breed and individual genetics — some goats just give poor-tasting milk).

Does housed in nasty, unsanitary conditions often give tainted milk, as do does kept with smelly bucks in rut. Clean the barn and move the buck; see if the flavor improves.

Some types of feed may flavor milk if fed within 5 hours of milking time. These include alfalfa and sweet clover (green or baled into hay), green barley, soybeans in any form, rye, rape, and vegetables such as turnips, cabbage, and kale.

Many plants, both wild and tame, impart unwelcome flavors. Our does nibble bitterweed for a few weeks each summer — yuck! Other things known to taint milk are wild garlic and onions, mustard, marigolds, chamomile, fennel, ragweed, peppercress, wild lettuce, and Queen Anne's lace.

Freezing Excess Milk

To freeze goat's milk, pour cooled, fresh-from-the-goat (refrigerated milk tends to separate when thawed after freezing) raw or newly pasteurized milk into shatterproof containers (we reuse 20-ounce [0.6 L] plastic water bottles; others prefer ziplock freezer bags), allowing 1 inch (2.5 cm) of unfilled air at the top for the frozen milk to expand into. Chill in icy water, then pop them in the freezer for up to 6 months. Thaw frozen milk slowly in the refrigerator to conserve its flavor and texture. Freezing affects taste a teensy bit, but it's still tastier than anything you can buy at the grocery store!

Goat's Milk Products

One milk goat may lead to two, two may lead to three, and before you know it you may have more milk than you can drink with breakfast, lunch, and dinner. Making cheese is a fun way to use extra milk (for some, it may be the major motivation behind buying a goat in the first place). And if you get really good at making cheese, you might even want to start selling it (more about selling later). For now, here are some basic cheese recipes. And if you're really adventurous, don't stop at cheese; goat's-milk yogurt is also great, and goat's milk–based soap is good for the skin.

· · · · · · · · · · · · · ·

Better a goat that gives milk than a cow that does not.

~ Estonian proverb

· · · · · · · · · · · · · ·

Easy Cheese: Queso Blanco!

· · · · · · · · · · · · ·

You can craft scores of yummy cheeses with goat's milk, but there is a learning curve. If you want to try your hand at cheese making right away, whip up a tasty batch of queso blanco.

Queso blanco ("white cheese," also called *panir* in India and *kesong puti* in the Philippines) is a rubbery white cheese with a bland, vaguely sweet flavor. It has the unique property of not melting when cooked (it can be browned or even deep-fried), and it sucks up the flavor of whatever it's cooked with. It's a terrific meat extender and tasty addition to chili, stir-fry and other Asian dishes, salads, and spaghetti. It can be flavored with herbs and spices as a between-meal nibble, and it freezes very nicely. What's not to like about queso blanco?

    1 gallon raw or pasteurized goat's milk
    ¼ cup white vinegar
      Salt and herbs (optional)

1. Warm the milk to 189°F (87.2°C), stirring with a whisk to prevent scorching. Maintain this heat for 10 minutes, then slowly add the vinegar until solid curds separate from the greenish whey.

2. Pour the mixture through a colander lined with fine cheesecloth (not gauzy cheesecloth from the craft store, which allows the mixture to squeeze through the weave). If you don't have real cheesecloth (available at cooking stores), a square of clean, cotton T-shirt material will work reasonably well.

3. Tie the corners of the cloth together and hang the bag to drain for 3 to 5 hours, until it stops dripping. This solid ball of curd can be stored in the refrigerator for up to 1 week. Or, if you prefer, crumble the ball and salt and flavor your cheese with herbs or spices. For a tangier version, substitute the juice of three to five lemons for white vinegar in the basic recipe. Delicious!

Yield: 1½ to 2 pounds

Chèvre

This recipe comes from *Home Cheese Making*, by Ricki Carroll. This easy, creamy cheese is another good one for beginners. Ricki says it can be used as a substitute for cream cheese or ricotta.

1 gallon pasteurized whole goat's milk
1 packet direct-set chèvre starter

1. Heat the milk to 86°F (30°C). Add the starter and stir to combine.
2. Cover and let set at a room temperature not below 72°F (22°C) for 12 hours. (Ricki says she likes to make the cheese at night, so she can drain it when she gets up in the morning.)
3. Line a colander with butter muslin. Gently ladle the curds into the colander. Tie the corners of the muslin into a knot and hang the bag over the sink to drain for 6 to 12 hours, or until the curds reach the desired consistency. (A shorter draining time produces a cheese spread; a longer draining time produces a cream cheese–type consistency.) A room temperature of at least 72°F will encourage proper draining.
4. Store in the refrigerator in a covered container for up to 1 week.

Yield: 1½ pounds (680 g)

Soft Goat Cheese

This recipe also comes from *Home Cheese Making* by Ricki Carroll. Ricki suggests that you may want to experiment with various sizes and shapes of soft-cheese molds from a cheese-making supply house. This cheese is a yummy spread for sandwiches, bagels, and crackers.

½ gallon pasteurized whole goat's milk
1 ounce prepared fresh starter
1 drop liquid rennet diluted in
   5 tablespoons cool, unchlorinated water
1 teaspoon cheese salt (optional)

1. Heat the milk to 76°F (24.5°C). Add the starter and stir to combine.
2. Add 1 tablespoon of the diluted rennet and stir with an up-and-down motion for 1 minute.
3. Cover and allow the milk to set for 12 to 18 hours, or until it coagulates. The room temperature should not exceed 72°F (22°C).
4. Scoop the curds into individual goat-cheese molds (approximately 3¼ inches [8 cm] in height). When the molds are full, put them in a convenient place to drain. Drain for 2 days.
5. After 2 days of draining, the cheese will have sunk to about 1 inch (2.5 cm) in height and will maintain a firm shape. Unmold the cheese, and, if desired, immediately lightly salt the surface to taste. Eat the fresh cheese now or wrap it in cheese wrap and store up to 2 weeks in the refrigerator.

Yield: About 1 pound (450 g)

Selling Goat's Milk

A few words of warning: Don't sell or give away goat's milk, especially raw milk, before you know that it's okay to do so. Selling milk from the farm is illegal in many states, and laws vary in the states that do allow it. In Arkansas, where we live, producers can sell a certain amount of raw goat's milk each month provided customers come to the farm to pick it up; raw goat's milk from home dairies cannot be sold in stores or anyplace else off the farm. Just a few miles north of us, Missourians can sell all they produce, not only from their farms but also through markets and even delivery routes! Stiff penalties for trafficking in illegal milk exist. Know your legal rights before you sell it.

You could also become a small-scale, licensed goat dairy. It's a heap of work but rewarding in so many ways. Keep in mind that all dairies, large and small, are licensed and regulated by their state's food safety administration. Before going this route, read "Start a Grade A or Grade B Goat Dairy" at the American Dairy Goat Association website (see Resources).

The Udder Truth

Your goat has two teats and one udder. The udder is her entire external mammary structure. You are not grasping her udders, as some folks mistakenly say; you are grasping only her teats.

THE CORN-SPIRIT AS A GOAT

When Sir James George Frazer wrote The Golden Bough: A Study in Myth and Religion *(first published in 1890), he devoted an entire chapter to the corn-spirit as a goat. In this case "corn" doesn't mean the tasty yellow vegetable we eat off the cob; it's the old European word for grain of all kinds. These are some excerpts.*

Further, the corn-spirit often appears in the form of a goat. In some parts of Prussia, when the corn bends before the wind, they say, "the Goats are chasing each other," "the wind is driving the Goats through the corn," "the Goats are browsing there," and they expect a very good harvest. Again they say, "the Oats-goat is sitting in the oats-field," "the Corn-goat is sitting in the rye-field." Children are warned not to go into the corn-fields to pluck the blue corn-flowers, or amongst the beans to pluck pods, because the Rye-goat, the Corn-goat, the Oats-goat, or the Bean-goat is sitting or lying there and will carry them away or kill them.

When a harvester is taken sick or lags behind his fellows at their work, they call out, "the Harvest-goat has pushed him," "he has been pushed by the Corn-goat." In the neighbourhood of Braunsberg (East Prussia) at binding the oats every harvester makes haste "lest the Corn-goat push him." At Oefoten, in Norway, each reaper has his allotted patch to reap. When a reaper in the middle has not finished reaping his piece after his neighbours have finished theirs, they say of him, "he remains on the island." And if the laggard is a man, they imitate the cry with which they call a he-goat; if a woman, the cry with which they call a she-goat.

At Gablingen, in Swabia, when the last field of oats upon a farm is being reaped, the reapers carve a goat out of wood. Ears of oats are inserted in its nostrils and mouth, and it is adorned with garlands of flowers. It is set up on the field and called the Oats-goat. When the reaping approaches an end, each reaper hastens to finish his piece first; he who is the last to finish gets the Oats-goat.

— Sir James Frazer, *The Golden Bough*

Shear Delight

*It's no use going to the goat's house
to look for wool.*

~ Irish proverb

If you would like to own a true multipurpose goat, investigate Angora and Cashmere fiber goats. Angora goats produce mohair and Cashmere goats produce cashmere. Smaller breeds, such as Pygoras and Nigoras, produce one, the other, or a combination of the two. Full-size Angora and Cashmere goats also excel in harness and as packgoats, and some give a respectable amount of milk.

Angora Goats

When most folks think of fiber goats, they think of Angoras. Angoras are curious and quiet, with endearing personalities, and they produce a lovely product: mohair. Quality mohair is in high demand with fiber crafters, who handspin the fiber as is or along with other fibers. They use mohair yarn for knitting, crocheting, and weaving; locks of Angora fiber for doll's wigs and Santa beards; and mohair

cord for crafting the highest-quality cinches for riding saddles.

Angora goats originated on the plains of central Anatolia in the area around present-day Ankara, Turkey, where they've flourished for at least 2,000 years. Historically they were small, refined goats with 8- to 10-inch-long (20–25 cm) lustrous, soft and silky, oil-free ringlets of fiber. They were covered in less fiber than are modern goats, however, and you could clip off about 4 pounds (1.8 kg) of mohair every year.

Angora goats arrived in the United States in the mid-1800s. Just prior to the American Civil War, Sultan Abdülmecid I asked President James K. Polk to send an expert to Turkey to help establish a cotton industry there. Polk sent Dr. James P. Davis, of Columbia, South Carolina. When the doctor returned to the United States, in 1849, he brought along seven Angora does and two bucks, gifts from the delighted

sultan. More importations to the United States followed from both Turkey and South Africa, so that by 1863, farmers in 17 states had herds of 12 to 300 Angora goats.

Although many herds were lost during the Civil War, the American mohair industry continued to thrive. During the late 1800s, most Angoras were raised in the farm states; around 1900 the tide changed and Angora ranching moved to Oregon and, even more so, to the Edwards Plateau region of west-central Texas.

Between 1940 and 1980, 95 percent of America's mohair was grown in Texas.

Today, South Africa, Australia, and New Zealand have thriving mohair industries, with South Africa leading the world in mohair production.

Angora Attributes

Angoras are medium-size, fine-boned goats; does weigh 80 to 110 pounds (36–50 kg) and males 125 to 200 pounds (57–91 kg). Angora

A modern white Angora and a colored Navajo Angora

goats are quieter than many other breeds and have softer voices, bucks are milder mannered and have less odor than other breeds, and Angoras tend to stay inside fences. They're ideal pets, especially for handspinners or anyone else with an interest in fiber arts.

The difference between goats registered in the American Angora Goat Breeders Association and those registered in the Colored Angora Goat Breeders Association is their color; everything else about them is the same. Goats registered by the Navajo Angora Goat Record differ from the others in fleece coverage, so let's look at that aspect first.

All Angoras grow mohair, but Navajo Angoras produce less fiber than do modern Angoras. That's because modern Angoras were selected for prolific fiber production; even the tops of their head, their cheeks, and their legs are carpeted with fleece. Original imports had an open face (the fleece line started behind the ears) and clean legs below the knees and hocks; that's how Navajo Angoras still look today.

The average white or colored Angora grows between 6 and 30 pounds (2.7 and 13.5 kg) of 12-inch (30 cm) fiber each year; they are shorn twice a year when locks are 4 to 6 inches (10–15 cm) long. Like their Turkish ancestors, Navajo Angoras grow only 3 to 4 pounds (1.4–1.8 kg) of mohair, but since it grows predominantly on prime areas, there is little waste when they are shorn. Because Angora fiber grows year-round, Angoras are shorn twice a year, usually 2 to 4 weeks prior to kidding and 6 months after that.

Angoras require excellent weather protection for 2 to 6 weeks after shearing, even in warm climates. Exposure to snow, rain, and cold weather contributes to respiratory diseases such as pneumonia. In addition to adequate shelter, it's wise to cover newly shorn Angoras with goat blankets or old sweaters and sweatshirts to help trap body heat. Another way to protect shorn Angoras is to "cape shear" them: leave a wide strip of unshorn fleece along each goat's back to protect him from chills.

Most Angora does produce single kids. Angora kids are more delicate than are other breeds; they need serious protection from chills in the form of a warm, draft-free indoor enclosure and lamb covers when it's frigid outside.

It takes a lot of protein to grow great fleece, so Angoras need a more precise diet than that of many other breeds. Again, consult your county Extension agent or a local, successful Angora goat breeder to formulate a diet based on the needs of fiber goats in your locale. Angoras, even wethers, usually require a basic diet of alfalfa hay or a grass hay- and grain-based diet, both supplemented with protein additives for kids and late-gestation and lactating does. As always, plenty of clean water and proper mineral supplementation are essential.

Angora Fiber

The luscious, lovely fiber Angoras produce is called mohair, from the Arabic word *mukhayyar,* a type of high-quality goat fiber cloth (*khayyara* means "select" or "choice"). Angora goats do not produce angora fiber; angora is harvested from Angora rabbits.

The diameter of a mohair fiber increases as an animal ages, so the finest grades come from kids. The finest grade — number one kid — is under 23 microns in diameter (1 micron is 0.00003937 of an inch); the coarsest grown —

number three — is 43.01 microns in diameter or more. Finer grades are used for making clothing; coarser ones are for making carpets and heavy outerwear.

Mohair is composed mostly of keratin; it has scales like wool but the scales are only partially developed, so mohair doesn't felt easily, in the manner that wool does. It is lustrous, nonflammable, stretchy, durable, and warm.

In Angora goat parlance, a "defect" is anything that becomes embedded in fiber prior to shearing. This can be vegetable matter like twigs and burrs, feed residue like hay chaff, or environmental junk like strings of fiber from the ubiquitous plastic tarps used on today's farms. A defect greatly devalues fleece, especially fleece marketed to handspinners.

Here are some tips for ways to protect your goats' fleece.

- Don't throw hay at feed bunks while goats are near enough to catch flying debris in their fleece.
- Feed Angoras at chin level or lower, so they don't reach up and pull hay onto their heads.
- Police pastures and paddocks for anything likely to get snagged on fiber.
- For the best fleeces, jacket your goats with a lightweight, reusable sheep cover (see Resources for names of suppliers). They make a dramatic difference in fleece quality, goats don't seem to mind them, and they last seven or eight years.

Shearing Your Angora Goat

Commercial sheep shearers often shear fiber goats when asked. If you can find one, consider yourself lucky, as he'll probably remove your goat's fiber in one nice piece. If you can't find a pro, however, it's reasonably easy to shear a goat yourself. Here's what you need to know.

You can clip your goat using heavy-duty horse clippers, sheep shears fitted with a mohair comb, or sharp scissors designed to be used by hand (Fiskars scissors work great).

To begin, place a tarp on the ground and set a fitting stand or milking stand on top of the tarp. Place the goat on the stand and, using a handheld blower (leaf blowers work fine), blow vegetable matter out of his fleece. Scissor the goat, taking care not to nick his sensitive skin, and sort clean body fleece into one pile and soiled bits into another. Most people like to start at the goat's back and work down, removing the best fleece first, though the order of shearing is strictly a personal preference.

Shearing is less traumatic for all concerned when your goat isn't frightened out of his wits! Begin early by clicker-training him to hop onto the stand and to allow his head to be fastened onto the chin cradle or his neck secured in the stanchion. Then, click and reward good behavior while you shear your goat. Next time, he'll be much happier to play along.

This amazing Angora goat was photographed at Lippe Studio in Del Rio, Texas. How long were his tresses? It's hard to say. Some Angoras of the era, whose fiber was kept in braids, grew tresses up to 40 inches (102 cm) long.

Sergeant Bill

ON AUGUST 23, 1914, a train carrying soldiers of the newly mustered 5th Western Cavalry Expeditionary Force stopped at Broadview, Saskatchewan, Canada. A group of recruits spied a young girl named Daisy Curwain and her cart goat, Bill. They asked Daisy if they could have her goat as their good-luck mascot, and she agreed.

Private Bill proceeded with his new family to Valcartier training camp and thence overseas to England. After long, arduous months of training, the unit received their orders to proceed to the front. However, they weren't allowed to bring any regimental pets.

The men of the 5th weren't ready to comply with that regulation. According to Sergeant Harold Baldwin, who penned *Holding the Line* (see Resources) while stationed on the front, "We could not part with Billy; the boys argued that we could easily get another colonel, but it was too far to the Rocky Mountains to get another goat. The difficulty was solved by buying a huge crate of oranges from a woman who was doing brisk trade with the boys. The oranges sold like hot cakes and in a jiffy the orange box was converted into a crate and Billy shanghaied into the crate and smuggled aboard the train."

Bill's life in the trenches was an exciting one. He was known for his fondness for canteen beer and his propensity to eat important papers left lying around. Bill redeemed himself in battle, though, earning the rank of sergeant. At Ypres he was found in a shell crater standing over a nervous Prussian guardsman, even though he himself had been wounded by shrapnel; at the Second Battle of Ypres he was gassed along with his boys. He fought at Vimy Ridge, was shell-shocked at Hill 70, and was wounded twice at Festubert, where he became a hero by knocking three soldiers into a trench seconds before a shell burst precisely where they had been standing. By war's end, he was one of the few original soldiers of the 5th still active and the only original mascot to enter Mons on Armistice Day. He was awarded the 1914/1915 Star, the British War Medal, and the Victory Medal for his 4½ years of service.

Sergeant Bill accompanied his men to Berlin at war's end and marched in the grand Victory Europe parade wearing an embroidered blue plush coat emblazoned with sergeant's stripes.

Despite immigration problems, Bill returned to Saskatchewan with his unit, where it was demobilized on April 24, 1919. He was later returned to Miss Cur-

wain in Winnipeg and lived several more years. After his death, Bill was stuffed, mounted, and displayed in the Saskatchewan Legislative Building. His body was eventually returned to his home in Broadview, where he still holds a place of honor in the Broadview Museum.

The pet goat of the Canadians

Senior mascot of the Canadian armies during World War I, Sergeant Bill accompanied the First Canadian contingent from Canada to the battlefields of Flanders and home again, earning the 1914 Star, the General Service Medal, and the Victory Medal along the way. This postcard was posted from Barry, Glamorgan, Wales, on January 26, 1916.

Washing Angora Fiber

Washing mohair is easy, easier than washing a comparable batch of wool. Just follow these simple steps.

1. Fill your washer with water that is 160°F (71°C). Place 4 or 5 pounds (1.8–2.3 kg) of defect-free fiber (handpick it before washing if it's full of junk) in a mesh bag, submerge it in the water, and let it soak for 30 minutes (don't swoosh it around in the water; just let it be).

2. Drain the water, and then use the washer's spin cycle to spin the fiber.

3. Refill the washer with 160°F water and stir in ¼ cup of detergent per pound of fiber. Submerge the bagged fiber and soak for 30 to 60 minutes. The water must stay hot; if it cools prematurely, drain some of the water and pour in additional hot water to bring it back to temperature. If the water gets too cold, grease in the fiber forms a scum that is difficult to remove.

Angoras as Explorers

Itinerant, turn-of-the-century photographers used fine, fancy Angora goats to power many a cart and wagon for a prop when photographing children. And at least one team of intrepid Angoras pulled a prospector's sled in the frozen Klondike. The caption on this photo, taken by Eric A. Hegg, reads, "Taken near Skagway, Spring 1898. A team of Angora goats, harnessed to a loaded sled. The owner's idea was probably that after serving as draught animals, they could become available as meat and their long-hair hides could be used as winter clothing or robes."

Nannyberries

TAD LINCOLN'S GOATS

Soon after President Abraham Lincoln's family moved to Washington, in 1861, Mr. Lincoln's son Tad asked for goats. Two goats, Nanny and Nanko, purchased for the grand sum of five dollars each, became the eight-year-old's "cohorts in crime."

Tad, always a high-spirited child, and his goats starred in a number of memorable White House escapades. On one occasion he harnessed them to his goat wagon and drove them pell-mell through the corridors of the East Wing. When the White House steward huffily reported this to Mr. Lincoln, the president smiled and inquired, "Well, are they outside again? Is Tad safe? How about the goats?"

The president also allowed Tad to take his goats up to bed with him at night, and when Tad drove them through a tea party held in the East Room, Mr. Lincoln hid himself in an adjoining room and roared with laughter.

According to Mrs. Lincoln's seamstress, Elizabeth Keckley, the president loved Tad's goats. In her memoir (*Behind the Scenes, or, Thirty Years a Slave and Four Years in the White House;* 1868), Mrs. Keckley wrote, "Mr. Lincoln was fond of pets. He had two goats who knew the sound of his voice, and when he called them they would come bounding to his side. In the warm bright days, he and Tad would sometimes play in the yard with these goats, for an hour at a time."

Mr. Lincoln and Tad visited the nearby Soldiers' Home almost daily, frequently taking along Nanny and Nanko in the presidential carriage. Eventually, to pacify Mrs. Lincoln (who wasn't amused by some of Tad's antics), the goats moved to the Soldiers' Home full time. That is, until Nanny so decimated the flower beds that she was sent back to the White House in disgrace — but only for a very short time. On August 8, 1863, Mr. Lincoln sent a telegram to his vacationing wife, saying: "Tell dear Tad, poor 'Nanny Goat' is lost; and Mrs. Cuthbert & I are in distress about it. The day you left Nanny was found resting herself, and chewing her little cud, on the middle of Tad's bed. But now she's gone! The gardener kept complaining that she destroyed the flowers, till it was concluded to bring her down to the White House. This was done, and the second day she had disappeared, and has not been heard of since. This is the last we know of poor 'Nanny'."

Nanny was apparently found, as on April 28, 1864, Lincoln sent a telegram to his wife in New York bidding her: "Tell Tad the goats and father are very well — especially the goats."

4. Drain and spin, then soak one more time in 160°F water that contains ¼ cup detergent per pound of fiber.

5. Rinse the bagged fiber by submerging it in detergent-free, 160°F water for 30 minutes, repeating until the rinse water is clear.

6. Dry the loose fiber on racks or screens outdoors (out of direct sun) or in a warm room indoors.

Marketing Angora Fiber

If you don't use your goat's fiber yourself, there is a ready market for clean Angora handspinners' fiber. Sell it on eBay or touch base with buyers through Angora goat and handspinner lists at YahooGroups (see Resources).

Angora Goat Registries

The American Angora Goat Breeders Association (AAGBA) (see Resources) is the oldest registry for Angora goats. It registers modern-type, white Angora goats based on pedigree (registrants must be the offspring of two AAGBA-registered parents).

In 1992, admirers incorporated the first colored Angora organization, the Colored Angora Goat Record. The Colored Angora Goat Breeders Association (CAGBA) (see Resources) followed in 1998. The CAGBA promotes and preserves the pedigrees of colored Angora goats (who are ineligible for AAGBA registration) but also registers white and lightly colored goats in a separate herd book. Registration is by pedigree (both parents must be registered with the CAGBA or the AAGBA) or by inspection.

The Navajo Angora Goat Record (NAGR) recently formed to preserve the pedigrees of rapidly disappearing primitive-type Angoras of the sort originally imported from Turkey. A remnant population still exists in parts of the American Southwest, particularly on the vast Navajo Reservation. Registration is currently based on fiber and type. Most Navajo Angoras are colored, but white ones are accepted too.

Cashmere Goats

Cashmere goats are not a breed per se; they're simply goats that grow an abundance of cashmere. Spanish goats are famous for producing fine cashmere, and one of my Boers produces several ounces of cashmere every year (a bona fide Cashmere goat yields roughly 2 to 4 ounces [60–120 g])! A type, however, is emerging, and the Eastern Cashmere Association (see Resources) publishes a comprehen-

---

Harvest *Your* Goat's Cashmere

Many goats grow enough cashmere to make combing them a productive venture. Check your goat's undercoat as late winter approaches. If it's thick, soft, and fluffy, begin combing as soon as the undercoat pulls loose. Continue combing for the next 2 weeks or longer, until no more undercoat comes out in the comb. Place the fiber in a bowl and carefully pick out stray guard hairs. Now you have some fine cashmere!

A beautiful, long-horned Cashmere buck

sive North American Cashmere Goat Breed Standard on its website.

Cashmere goats are hardy, easy-care goats with oodles of personality. While not as laid-back as Angoras, the big, strong, beautifully horned goat described in the North American Cashmere Goat Breed Standard makes a fine harness and packgoat, as well as a fine fiber producer. And the beauty is that you needn't shear these goats to harvest their fiber; simply comb them when they shed their winter hair. Cashmere sheds first, so very little guard hair comes off in the comb.

Most goats have soft insulating hairs near their skin and long guard hairs over that. Guard hairs are of little value to crafters, as they are too coarse, hard to spin, and difficult to dye. The undercoat, however, is cashmere. By industry standards, cashmere fiber must have an average diameter of less than 19 microns and be at least 1¼ inches (3 cm) long. Its crimpiness gives it loft, providing warmth without weight. The only goats that don't grow cashmere are single-coated goats such as Angoras.

Cashmere goats can be any color, though their cashmere-producing neck and body should be the same solid color. Guard hair can be long or short but must be coarse enough to be easily distinguished from cashmere. Horn styles vary, but bucks and wethers tend to grow huge, impressive horns.

Nigoras and Pygoras

Nigoras are created by crossing Nigerian Dwarf goats with full-size Angoras; Pygoras are created by crossing Pygmy bucks with full-size Angora does. Both breeds produce three types of fiber that vary by individual: some produce mohair, some cashmere, and many fall someplace in between. Mating two animals of one type of fiber will not guarantee kids of the same type. Both breeds are small but very productive fiber goats that are also fun, friendly pets.

Pygora

Nigora

William de Goat

Arguably the most famous goat to serve in World War II was Air Commodore William de Goat DSO DFC, mascot of the British 609 West Riding Squadron, who authored a book about his wartime experiences (with the help of his friend Squadron Leader Brian Waite, RAF, of course).

William, a handsome British Toggenburg kid, joined the squadron at Biggin Hill, near London, on June 23, 1943, and was immediately commissioned a flying officer. He remained with the squadron until its disbandment in 1945, when he retired with the lofty rank of air commodore.

During those two years, with the help of "his boys," he sniffed oxygen bottles and consumed innumerable cigarettes, along with top secret maps, files, and records. He flew with his squadron on several occasions, including to Europe, where he disembarked at Normandy and subsequently served at 12 airbases. He was shot at but never hit, though he bloated badly after munching mattress stuffing meant to be bedding.

At one point he even almost died of slow poisoning from licking paint. That occurred because as William rose through the ranks, his boys used blue "aircraft dope" to paint bars denoting his rank on his horns. William, who liked the taste, would swipe his horns across his front legs and lick off the paint. Just prior to deployment to Europe, he became quite ill, whereupon the airbase's medical officer determined that the blue paint was toxic. After that, William's horns remained unadorned.

If you love goats, the charming little book of William's exploits, *William de Goat* (see Resources), is a must-read. Check it out. I think you'll be glad you did.

Part 3

Caring for Your Goats

9 Housing, Equipping, and Feeding Goats

*If you go into a goat stable, bleat; if you
go into a water buffalo stable, bellow.*

~ Indonesian proverb

Goats are an easy-care species. You don't need a lot of equipment to keep goats. There are a few essentials: housing, pens, fences, everyday items such as halters and leads, and, of course, feed.

Simple Shelters

Goats need a draft-free, dry place to get out of the sun and weather. That's it — the essence of simplicity.

An adult goat of any full-size breed will do well with 15 to 20 square feet or so (1.4–1.9 sq m) of living space under shelter, and if he's not pastured every day, he'll need another 25 square feet (2.3 sq m) of exercise area outdoors. Midsize and miniature breeds need correspondingly less room. Goats with full-time access to pasture or an exercise run can crowd closer together indoors during inclement weather — but not for long. Crowding leads to stress and stress to aggressive behavior and disease. Give your goats room to breathe.

Goats thrive in refurbished chicken coops, homemade A-frames, extra-large doghouses, calf hutches, palatial goat barns, or stalls in the horse barn. Some of our goats live in homemade, goat-size loafing sheds, and the rest have Quonset-style Port-a-Huts (see Resources). If we could start over, we'd use only Port-a-Huts. The goats love to climb on them (be sure to secure the flap in the back wall so goats' legs don't get caught and injured while climbing), they're inexpensive yet tremendously sturdy, and cleanup is as easy as picking up the hut and hauling it to a new location. Whatever you choose, make sure you can move it, get inside it, or take it apart to clean it.

The most important attributes of any goat shelter are a draft-free sleeping area and proper ventilation. Goats kept in a drafty structure or poorly ventilated, enclosed winter housing are prone to respiratory ailments such as pneumonia. Don't shut all the windows and doors. Goats can handle subzero weather as long as they're dry. If your goat gets cold, buy (or make) him a goat coat; don't seal him up in an airtight barn.

Build your structure with solid wood whenever you can. Don't expose your goats to composite building materials such as particleboard and pressed board; even lightweight plywood is suspect. Goats love the taste and will eat it with relish. Formaldehyde is commonly used in these manufactured wood products because of its preservative and adhesive properties. Formaldehyde is a known carcinogen and an allergen that can irritate eyes, mucous membranes, and the upper respiratory tract. Chewing wood also generates dangerous splinters that can lacerate a goat's gut.

Floors can be made of concrete, stone, packed dirt, sand, or clay and bedded with straw, rice or peanut hulls, dust-free shavings, or rubber stall mats. Many folks build sleeping platforms for their goats. Some use rebounder trampolines from the used-a-bit store (make sure spaces between springs are wide enough not to catch a goat's legs); others like suspended canvas dog beds. Our platforms are made of heavy exterior plywood fastened to free wooden pallets from our feed store. We cover them with straw for bedding and also use them for flooring in Port-a-Huts.

Coats for Goats

You can purchase horse-style winter blankets for goats of all sizes, though you can make great goat covers for much less money.

Visit the Goodwill store or shop garage sales to find secondhand sweatshirts and cardigan sweaters sized to fit goats (wool cardigans are the crème de la crème of caprine couture). Trim off the sleeves, maneuver the goat's forelegs through the armholes, and button the sweater up his back. There you are!

Dog sweaters make cute, inexpensive kid coats. Don't be afraid to buy pullover styles just because it doesn't look as if they'll fit; they stretch a lot farther than you'd think. Or snip neck and leg holes in a man's extra-large winter wool sock to fashion a snug-fitting coat for a preemie.

This is our goat Tallulah perched atop one of our Port-a-Huts. Not all goats can climb them, but some do.

Try to Pen Them In

If you build goat quarters inside a larger structure, you'll need pens to contain your goats. Make them tall; some goats leap like gazelles. Enclose them with solid partitions anyplace drafts are an issue. Or try welded-wire cattle or sheep panels for great ventilation during the summer months; during the winter, cover the panels with plastic tarps. They aren't elegant but they work!

Cattle panels are my material of choice for building pens, exercise runs, and outdoor paddocks. Cattle panels, sometimes called stock panels, are prefabricated lengths of mesh fence welded out of galvanized ¼-inch steel rods. Most cattle panels are 52 inches (132 cm) tall and built using 8-inch (20 cm) stays; horizontal wires are set closer together near the bottom of the panel to prevent small kids from

wriggling through. They're sold in 16-foot (5 m) lengths that can be easily trimmed to size using heavy-duty bolt cutters.

Sheep panels are cattle panels manufactured in 34- and 40-inch (86 and 102 cm) heights. Their horizontal wires are set even closer together; they're good for stacking to achieve additional height.

Utility panels are another option. They're 20 feet (6 m) long, welded out of extra-heavy-duty 4- or 6-gauge rod and fabricated using 4-inch by 4-inch (10 cm × 10 cm) spacing. They come in 4- to 6-foot (1.2–1.8 m) heights and are ideal for building buck pens.

The downside of cattle, sheep, and utility panel construction is that the raw end of each rod is very sharp. To make these panels more user-friendly, smooth the end of each rod with a rasp to zip off its razor edge

Premier1 Supplies (see Resources) sells panels designed for goat and sheep applications, and their rod ends are presmoothed at the factory. They come in 36- and 40-inch (91 and 102 cm) heights and in 4- to 6-foot (1.2–1.8 m) lengths — just right for building V-type hay bunks. Premier1's panels or cut-to-size standard cattle panels are also used for walk-through gates, kidding jugs, and small-size temporary pens.

Build the Best Fence

You can't keep goats in poor fencing. It's hard to keep goats in *good* fencing. Probably the hardest, most aggravating aspect of goat keeping is keeping your goats where they belong. This is because goats have an adventurous spirit; they want to see what's beyond the next hill. They also want to tap-dance on the hood of your car and nibble your neighbor's roses, never mind the garden in your own backyard. Don't scrimp on fences. Better a small pasture with sturdy fences that are tall and stout enough to keep goats in (and predators such as dogs and coyotes out) than a large one ringed with questionable fencing.

Several types of farm fences work well with goats. Cattle panels are an obvious first choice. They are strong, durable, and easy to erect: set wooden or steel fence posts, carry the rigid panels into place, wire them to steel posts or tack them to the inside of wooden ones, and there you are. Unfortunately, enclosing large areas with panels is costly. A close second best is quality woven wire.

Woven wire, also called wire mesh or field fence, is made of horizontal strands of smooth wire held apart by vertical wires called stays. The horizontal wires are usually set closer together near the bottom of the fence. The vertical stays in standard woven wire are 6 inches (15 cm) apart, and horned goats tend to get their head stuck between them. It's dangerous. A trapped goat is at the mercy of his herd

This goat looks silly, but a bar taped to her horns keeps her from getting her head caught in a wire fence.

mates, predators, the elements, and, unless you free him promptly, himself (stuck for a long time, frightened goats go into shock, their systems shut down, and they die). Woven wire designed for goats has 12-inch (30 cm) stays, which are usually long enough to avoid this problem. If your adult goat continually catches his head in the fence, duct-tape a single length of small-diameter PVC pipe or wood dowel to the tip of both horns. It looks silly but it works. Give him a month or so to forget why the fence was so alluring, then remove the apparatus and see if he has reformed.

Correctly installed, woven wire is the most secure type of affordable goat fencing, making it perfect for perimeter or boundary fences. Four-foot woven wire contains most goats. Installing one or two strands of electric wire above the woven wire helps keep predators out, too.

The downside: Goats lean into wire-mesh fences and then stroll from post to post to scrub their sides. It's a great way to shed winter hair or a vexing fly, but it's awfully hard on the fence. To prevent this, string a strand of offset electric wire on the inside of the woven wire right at a goat's shoulder height. It works!

The Lowdown on Everyday Equipment

You'll also need everyday items such as halters, collars, leads, feeders, watering devices, and maybe a way to haul your goats. Here's what you should know.

Halters, Collars, and Leads

You'll need a means of leading and restraining your goat. Some folks like halters, some swear by collars; chances are you need both. We use collars on a daily basis but prefer halters for when we need to have greater control, such as on the trail or leading a goat in a parade. Halters easily snag on things, so they shouldn't be kept on goats full time.

Goat halters are the same type as used on alpacas and llamas. They're usually crafted of washable nylon strapping. The best halters we've used are strong and infinitely adjustable, and have solid brass fittings; they're designed by and available from llama gurus Cathy Spaulding and Marty McGee Bennett (see Resources). When buying halters, look for strong stitching and sturdy hardware. Nylon is almost unbreakable; stitching and fittings aren't. Be sure the halter you buy fits your goat's head. It shouldn't squash his face, but it shouldn't be too loose either.

Some people keep collars on their goats all the time. A collar provides a handy handle when you need it, but it can also be a safety

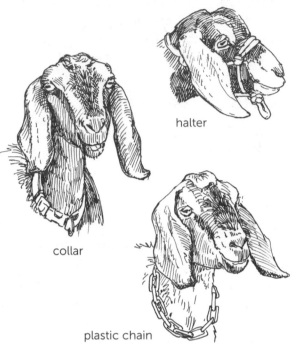

halter

collar

plastic chain

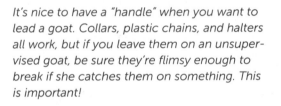

It's nice to have a "handle" when you want to lead a goat. Collars, plastic chains, and halters all work, but if you leave them on an unsupervised goat, be sure they're flimsy enough to break if she catches them on something. This is important!

hazard. Don't use strong, double-ply collars for everyday use — if an everyday collar snags on something and it hangs your goat, you want it to break! We like plastic chain collars from Hoegger Goat Supply (see Resources) and cheap, single-ply nylon dog collars from the local dollar store. Plastic chain comes in two link sizes and a dazzling array of colors. This type of collar fastens with a plastic connector that is designed to pull apart if the collar gets snagged and the goat pulls back hard; just pick it up and put the collar back on the goat. The cheap metal or plastic fittings on dollar-store

Nannyberries

dog collars break under pressure; they aren't reusable, but at a dollar apiece, they're cheap life insurance for your goat.

Leads can be anything from dog leashes to the sort of ropes used to lead horses. Choose something that complements your goat: a dog leash for a cute Nigerian Dwarf but a sturdy horse lead for a big Boer wether. Again, hardware is everything. Cheap pot-metal fittings will let you down. For safe leads that last, choose brass, nickel-plated brass, or stainless-steel fittings.

Feeders

All good goat-feeding equipment has one thing in common: it's designed to discourage goats from wasting feed. Goats, being the fastidious creatures they are, won't touch hay, grain, or minerals they or any other animal has peed or pooped in; conversely, they don't think twice about peeing and pooping in their own feed. Kids nest in accessible hay racks and grain or mineral feeders, and they don't vacate their nests when nature calls. It's

Combination hay and grain feeders give goats a second chance to eat clean, dropped hay — a good way to save on feed dollars.

Keyhole feeders save hay but are a danger to goats that use them.

your mission to prevent these unsanitary and wasteful practices.

Don't feed hay or grain directly off the ground. Doing so increases the likelihood of parasitism and disease, not to mention that it wastes feed.

Consider using grain feeders you can hang on the fence and remove after your goats have eaten, so they won't get pooped in. Or mount grain feeders 6 inches (15 cm) higher than your tallest goat's tail and provide booster blocks or rails for your goats' front feet to stand on.

Invest in a combination hay rack and tub-type feeder so that "wasted" hay falls in the tub, giving goats a second chance to eat it. Allow 16 to 18 inches (40.5–46 cm) of feeder space per average horned goat and 12 inches (30 cm) if your goats are disbudded or polled. The feeder should be large enough to allow all goats to eat at the same time; otherwise,

timid goats won't get their portion. This kind of feeder is also easy to keep clean.

Or make your own fence-line hay feeder by wiring the bottom of 4-inch- × 4-inch-mesh welded-wire panels to an existing fence and adding sturdy wire spreader arms at the top.

What you shouldn't use, no matter how much some goat owners praise them, are keyhole feeders. These are solid-sided wooden feeders with keyhole-shaped openings just big enough for a hornless goat to reach through to get some hay. They do save feed, but a goat with her head through an opening is at the mercy of her fellow diners. She can't see an attacking goat in time to pull back. Rammed hard in the side, she could die. Better some wasted feed than a dead goat.

Whatever style you choose, V-shaped hay feeders with welded-wire sides or vertical or diagonal slats work better than models with horizontal slats or bars.

Watering Devices

Goats drink ½ to 3 gallons (1.9–11 L) of water a day, depending on weather conditions (goats require more water during the hot summer months). Lactating does have the highest requirements, and to prevent urinary calculi, it's important for male goats to drink a lot, too. When water is contaminated with goat droppings, algae, dead birds or bugs, leaves, and other debris, goats drink only enough fluid to survive. Dump questionable water on a daily basis, and scrub buckets and tubs before refilling. And choose a series of small containers over one big tank; smaller ones are easier to clean.

During the summer months, place water tubs and buckets in the shade. This helps inhibit algal growth and because the water stays fresher, the goats will drink a lot more.

When it's *really* hot, freeze water in plastic milk bottles and submerge one in each trough or tub around noon; they'll cool water for several hours during the heat of the day and your goats will appreciate this treat. Refreeze them overnight and they'll be ready to use again the next day.

During the winter, keep water supplies from freezing with bucket or stock-tank heaters. Be sure to sheathe the cords in PVC pipe or garden hose split down the side and taped back together with duct tape; goats have been known to gnaw through electrical cords and die.

Where kids are present, shallow is the rule. Kids can leap or tumble into buckets or tubs and drown. *Never* use 5-gallon recycled plastic food-service buckets or any other narrow, deep water containers in kidding pens. It's always better to use two shallow water containers than one big, deep one when they'll be used by kids.

A Goat Conveyance

Goats can be hauled in horse trailers, goat totes (a type of goat-specific cage that slides into the bed of a pickup truck), truck toppers, mini-vans, and SUVs. You can use just about anything — as long as it's big enough, properly ventilated, provides safe footing, and is escape-proof.

Being escape-proof is important. Frightened or persistent goats can leap higher and squeeze through smaller openings than you can imagine. Err on the side of caution. If your goat bails out while you're driving down the highway, he's sure to be killed.

Restrain goats hauled in vans and SUVs; you don't want one jumping into your lap.

Thrifty Ways to Feed and Water

Need an inexpensive way to feed hay to one or two goats? Use a hay bag designed for feeding horses — not a hay net, but a *bag* made of solid fabric with an opening in the side to access hay. Goats catch their legs in hay nets; don't use them.

If you have cattle, you probably have empty plastic mineral lick tubs sitting around. Smaller ones make first-rate feeders and kid-watering troughs. Big ones are handy for watering adult goats.

Extra-large airline crates for dogs are perfect for hauling miniature breeds and kids. Secure larger goats with halters and leads; they travel best facing the rear of the vehicle or sideways, just like horses.

Giving Grub to Goats

Goats' nutritional needs depend on a slew of variables, including their age, sex, level of productivity (dairy doe, idle pet, hardworking wether), and the types of feed (especially forage) available where you live.

I recommend a diet that's based on forage (browse, pasture, or hay) augmented by a goat-specific mineral supplement formulated for the type of forage you feed and the area where you live, as well as a judicious amount of commercially bagged grain. For specifics, touch base with your county Extension agent and discuss a diet based on local feeds.

Your goat's digestive tract is designed to digest the cellulose in forage. It doesn't adapt well to starchy grain diets, which predispose goats to serious nutritional diseases such as bloat, acidosis, enterotoxemia, goat polio, listeriosis, laminitis, and urinary calculi. And some types of goats, such as dry does and wethers, thrive on a diet of forage alone. For these goats, mineral supplements can be used to balance the ration without exposing them to the dangers of grain overconsumption.

A Peek at Your Goat's Digestive System

All true ruminants (goats, sheep, cows, deer, bison, giraffes, and the like) have "stomachs" made up of four chambers: the rumen, reticulum, omasum, and abomasum.

The rumen. This is the first and largest of the four chambers. It's essentially a roomy fermentation vat. As newly consumed feed mixed with saliva flows into the rumen, it separates into layers of solid and liquid material. Later, when your goat is resting, he regurgitates a glob of food (his "cud") and rechews the material more slowly, then swallows it again. During second (and subsequent) chewings, feed is mixed with more and more saliva. Saliva, being alkaline, helps regulate rumen pH at a healthy 7.0 to 7.8.

The reticulum. Once liquefied, macerated feed leaves the rumen through an overflow connection called the rumino-reticular fold and enters the reticulum, where fermentation continues (methane is produced by bacteria and protozoa in the rumen and reticulum; this

Battling Bloat

Bloat occurs when a goat overeats grain, legume hay, or tender, high-moisture spring grass. Gas becomes trapped in his rumen and expands until it presses so hard against his diaphragm that he suffocates. A bloated goat's sides will bulge to an alarming degree; he'll also kick at his abdomen, grunt, cry out in pain, or grind his teeth.

If you suspect a goat is bloated, call your vet without delay. To prevent this life-threatening situation, store grain and legume hay where your goats can't overindulge. It's also wise to feed grass hay in the morning, before turning goats onto lush, spring pasture.

is the source of the stinky burps goats emit). Then the feed passes through a short tunnel into the omasum.

The omasum. This section is divided by long folds of tissue that resemble the pages of a book and are sometimes referred to as leaves. These leaves are covered with teensy, fingerlike structures called papillae, which help decrease the size of feed particles, remove excess fluid, and absorb any volatile fatty acids that weren't absorbed in the rumen.

The abomasum. The fourth chamber, the abomasum, is considered the goat's true stomach because it's where digestion occurs, much as it does in the human stomach. The abomasum, like the omasum, is lined with many folds that increase its surface area many times over. The walls of the abomasum secrete digestive enzymes and hydrochloric acid, which quickly lower the pH of feed slurry from about 6.0 to around 2.5. Protein is partially broken down in the abomasum before slurry is shunted to the small intestine.

The small and large intestines. As semi-digested feed flows into the small intestine, it's mixed with secretions from the liver and pancreas that push the pH back up from 2.5 to between 7.0 and 8.0. The higher pH makes it possible for the enzymes in the small intestine to reduce remaining proteins into amino acids, starch into glucose, and complex fats into fatty acids. As this occurs, muscular contractions continually push small amounts of material through the system and into the large intestine, where bacteria finish digesting the mix.

Your goat's rumen contains billions of bacteria, protozoa, and other microbes that feed on the carbohydrates in his diet and convert them to volatile fatty acids (his primary source of energy); without them, he would die.

One type of ruminal microorganism digests cell-wall carbohydrates (cellulose, hemicellulose, pectin, and lignin: in other words, fiber); these do best in a neutral environment, around 7.0 pH. Another type digests cell-soluble carbohydrates (sugars and starches found in

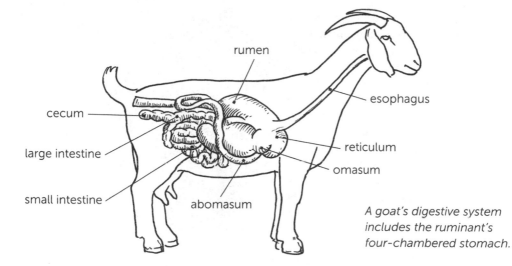

A goat's digestive system includes the ruminant's four-chambered stomach.

Cadwaladr's Goat

CADWALADR owned a very handsome goat named Jenny, of whom he was extremely fond; and who seemed equally fond of him; but one day, as if the very diawi possessed her, she ran away into the hills, with Cadwaladr tearing after her, half mad with anger and affright. At last his Welsh blood got so hot, as the goat eluded him again and again, that he flung a stone at her, which knocked her over a precipice, and she fell bleating to her doom.

Cadwaladr made his way to the foot of the crag; the goat was dying, but not dead, and licked his hand — which so affected the poor man that he burst into tears, and sitting on the ground took the goat's head on his arm. The moon rose, and still he sat there. Presently he found that the goat had become transformed to a beautiful young woman, whose brown eyes, as her head lay on his arm, looked into his in a very disturbing way. "Ah, Cadwaladr," said she, "have I at last found you?'"

Now Cadwaladr had a wife at home and was much discomfited by this singular circumstance; but when the goat — yn awr maiden — arose, and putting her black slipper on the end of a moonbeam, held out her hand to him, he put his hand in hers and went with her. As for the hand, though it looked so fair, it felt just like a hoof.

They were soon on the top of the highest mountain in Wales, and surrounded by a vapoury company of goats with shadowy horns. These raised a most unearthly bleating about his ears. One, which seemed to be the king, had a voice that sounded above the din as the castle bells of Carmarthen used to do long ago above all the other bells in the town. This one rushed at Cadwaladr and butting him in the stomach sent him toppling over a crag as he had sent his poor nanny goat.

When he came to himself, after his fall, the morning sun was shining on him and the birds were singing over his head. But he saw no more of either his goat or the fairy she had turned into, from that time to his death.

— *British Goblins: Welsh Folk-lore, Fairy Mythology, Legends and Traditions*, Wirt Sikes

concentrates such as grains and commercial bagged feeds); these prefer a more acidic environment, in the 5.0 to 6.0 pH range.

What your goat eats affects the pH of his rumen. When he eats fibrous forage, he cuds more often and releases more saliva; goat saliva contains bicarbonate to buffer the rumen, thus preventing it from becoming too acidic. When he eats concentrates, which have a low fiber content, less chewing is required so less saliva is produced. Also, the starch in grain is rapidly fermented, resulting in overabundant ruminal acid production. When your goat eats too much grain, his rumen's pH plummets and forage-digesting microorganisms die off, clearing the way for lactic acid–producing microbes to proliferate and produce more acid. The result: acidosis (see the box on this page).

In short, your goat's digestive system needs long-stem fiber to operate optimally. Some goats, especially growing kids and late-gestation or lactating does, do need grain in their diets due to their increased energy and protein needs. This is where bagged concentrates shine.

Hay It Is

Regardless of whether or not you give grain to your goat, he needs good hay, at least when he isn't out on pasture.

Given a choice between small "square" bales (they're actually rectangular) and big rounds, choose the small square bales. They separate naturally into "flakes," so you can dole out exactly what your goats need. Large round bales are okay if they've been stored under cover and are free of mold, but your goat will pull down more hay than he eats, then poop and lie in it. If he's a fiber goat, you also can expect a lot of debris-contaminated fleece at shearing time.

If you can, buy nutrient-tested hay from a reputable hay dealer. Be extra cautious if you buy hay at auction. Some sellers pile their best bales on the outside of the stack and tuck junky hay deep inside, out of sight.

Inspect the hay you plan to buy. Pay for two or three bales you can open. See what they look like inside; if the hay doesn't pass muster, you aren't obligated to buy any more.

Be Aware of Acidosis

Acidosis, also called lactic acidosis or grain overload, occurs when the rumen pH falls below 5.5. As the pH continues to decline, rumen microbes die and ruminal action decreases or ceases. Symptoms are depression, dehydration, bloat, racing pulse and respiration, staggering, coma, and death; survivors generally fail to thrive due to permanent damage to the lining of their rumen and intestines.

If you think your goat has acidosis, call your vet immediately! To prevent it, goats should be fed mainly forage, not grain. Also be sure to make all feed changes gradually (see page 150), thus allowing rumen microbes time to adapt to a different diet. Goats consuming high levels of concentrates should have access to a container of sodium bicarbonate (baking soda; buy it in bulk at the feed store) that they can nibble at as the need arises.

HOW TO EVALUATE HAY

GOOD HAY	BAD HAY
• Is green. Quality alfalfa is dark green. Good grass hay is light to medium green. The outsides of both legume (alfalfa, clover, lespedeza) and grass (timothy, brome, orchard, Bermuda) hays become bleached if exposed to light in storage, but inside the bales the hay should still be green.	• Is pale to golden yellow inside and out. This hay was sun bleached in the field. Its nutrition is poor and it is apt to be dusty. It's not a good buy.
• Doesn't have spots or dust and smells fine.	• Has white dust, blemishes, or black spots in hay, or has a musty smell — this hay was cut and stored before it was fully dry. Even if you don't spot actual mold, don't feed dusty hay.
• Has thin, flexible stems that easily bend without breaking. Quality hay comprises leafy plants mown just before blooming or sometimes in earliest bloom.	• Is coarse and stemmy. It is low in nutrients, and goats don't like it; most of the plant's protein is in its leaves, whereas stems are mainly low-energy cellulose.
• Is free of debris.	• Contains sticks, rocks, dried leaves, weeds, and insect or animal parts. Your goat can't eat sticks and rocks, the weeds might be toxic and their seeds will infest your property, and baled insects and desiccated animals cause serious disease.
• If fresh from the field, is aged for 5 to 6 weeks before feeding. If old, it has been carefully stored.	• Is uncured and from a new crop. Uncured hay can cause serious bloat.

Buy only the amount of hay you can store under cover. Stack it on pallets, old tires, or poles to get it off the floor; if you don't, the bottom layer will rot. Don't let barn cats use your haystack for a litter box or have kittens in it — cats are primary carriers of toxoplasmosis. Finally, to minimize waste and prevent disease, use those feeders. Never feed valuable hay from the ground.

Minerals Are a Must

Your goat should have access to a mineral supplement formulated specifically for goats. The most important minerals in goat rations are salt (sodium and chlorine), calcium, and phosphorus. Calcium and phosphorus are necessary for proper growth and development, but high levels of phosphorus in relation to calcium can cause urinary calculi.

Urinary calculi (also called bladder or kidney stones) are hard lumps of mineral crystals that form in the urinary tracts of sheep and goats. Does generate urinary calculi, but because of their shorter, straighter urethra (the tube that empties urine from the bladder), they can pass these stones with little or no discomfort. Wethers and bucks aren't so lucky. Stones easily lodge in a male's longer, skinnier urethra, especially where the small-diameter, wormlike urethral process (colloquially known as the pizzle) extends beyond his penis. When this happens, he can't urinate (in partial blockages he might be able to dribble) and unless the condition is corrected (and quickly), uremic poisoning sets in, his bladder or urethra ruptures, and he dies.

A goat suffering from a blockage caused by urinary calculi will show at least some of these symptoms:

- Difficult, painful, and/or dribbling urination
- Blood in his urine
- Straining, abdominal contractions, tail twitching, kicking at his abdomen
- Crying out or groaning in pain
- Standing in a rocking-horse stance with his front legs perpendicular to the ground and his hind legs held at an angle behind him
- Crystals in the hairs around his prepuce
- A swollen or distended penis; touching it makes him cry out in pain
- Lack of appetite, depression, collapse, death

If you think your male goat has a stone, call your vet without delay! Treatments are:

- Surgical removal of the urethral process. This restores urine flow in about one-third of early cases.
- Administration of acepromazine maleate (Ace), a prescription animal tranquilizer, or flunixin meglumine (Banamine), an anti-inflammatory painkiller. Either of these may calm spasms of the urethra and relax the muscles surrounding the penis long enough for your goat to pass the stone.
- For minor blockages, a 2-week course of daily doses of 2 tablespoons of ammonium chloride dissolved in 60 cc of water sometimes works.
- Anesthetizing the goat and tapping his bladder to remove urine buildup. This isn't a cure, however — it's a stop-gap procedure to buy more time.
- Surgery to drain the bladder and remove urinary stones.

A goat that successfully passes one stone is at risk to make more. The stone should be taken to a vet, who can submit it to a testing laboratory to determine which of several mineral combinations it's made of. Then the vet can formulate a treatment and prevention plan for all the male goats on your farm.

Here are some things you can do to reduce the instance of urinary stones.

Formulate rations so they provide a 2:1 calcium-to-phosphorus ratio (2 parts calcium to 1 part phosphorus). Most bucks and wethers who develop urinary calculi are on high-grain, low-forage diets or diets in which legume forage, especially alfalfa, predominates. It's important to feed bucks and wethers a ration based primarily on browse, good grass hay, or similar forage and a mineral supplement designed to balance the diet.

Provide lots of clean water. Goats who drink enough water will have a healthy urine flow; when wethers and bucks don't drink enough water, their urine becomes concentrated and crystals may start to form. Male goats need constant access to a fresh, clean supply of drinking water kept cool in the summertime, liquid in the winter (use a heated bucket or warm it with hot tap water several times a day), and inside the goat shelter during periods of inclement weather (goats won't troop to an outside water trough if they have to get wet). And keep it clean — if you wouldn't drink from your goats' water source, chances are they won't either.

Add loose salt to your goats' diet at the rate of 3 to 4 percent of their total grain ration. This encourages them to drink more water.

Add ammonium chloride to your male goats' diet to acidify their urine. Ammonium chloride makes crystal components more soluble, so they're more likely to be expelled in the urine before they form stones. Add preventive doses of ammonium chloride to grain rations at the rate of 1 level teaspoonful per 150 pounds of goat, divided into two daily feedings. Ammonium chloride tastes incredibly nasty, so you may have to conceal it in a blob of honey, molasses, jam, or flavored yogurt.

Give future pet and working wethers' urethras time to semi-mature by castrating them at 4 to 6 months of age. When you castrate kids at a young age, you remove the hormonal influence on the full development of the urinary tract.

Finally, Switch Feeds Gradually

If your goat's rumen doesn't contain the necessary microorganisms to digest a certain food, he can't digest that food. Don't make fast changes in feed type or content (and that includes going from hay to fresh spring grass or browse). Mix the diet over a period of 7 to 10 days: some of the old, some of the new, gradually phasing out the old. If you don't, good microbes die and others proliferate, sometimes leading to enterotoxemia (see chapter 10, page 168).

Your Healthy Goats

*When you are sick you promise a goat,
but when you are well again
make do with a chicken.*

~ Nigerian proverb

You've bought yourself a healthy goat and want to keep him that way. It shouldn't be difficult; goats are naturally hardy beasts, when they're properly cared for. Here's what you need to know to keep your goat in the pink.

The Health and Wellness Basics

Your first task is to know when your goat is ailing (turn back to chapter 2 and reread "How to Choose a Healthy Goat"). Then follow these tips.

Check on your goats at least twice a day, even if they're out on pasture. Make certain all goats are accounted for (sick goats sometimes drift away from the herd).

If something is wrong, address the problem right away; don't wait to see if it gets bet-

ter by itself. If you don't know what's wrong with your goat or you're not positive you know how to treat it, *get help*.

Police your pastures and exercise yards on a daily or weekly basis. Check for broken fences, sharp edges on metal buildings, protruding nails, stray hay strings, hornets' nests, and anything else a goat could get hurt on.

Don't allow goats to climb on stored hay; they can slip between the bales and get wedged in, injuring themselves or even suffocating. (Not to mention they will contaminate the stacked hay with feces and urine.)

Keep your goats in clean, dry surroundings. Provide draft-free shelter from the elements. Avoid overcrowding, as this leads to stress.

Choose goat-friendly feed. Feed quality hay; don't feed dusty or moldy bales. If you don't know how to formulate nourishing goat grain combinations, feed bagged goat concentrates. Make feed changes gradually, giving goats' rumens time to adjust to new types or quantities of feed.

Provide an unlimited supply of fresh, clean water; you want goats (especially bucks, wethers, and lactating does) to drink as much water as they can.

Keep mice, rats, opossums, and the like out of feed storage areas. Don't let cats, dogs, or poultry foul your hay.

Provide predator protection of some sort, be it exceptionally good fences or a livestock guardian dog, donkey, or llama. Thousands of goats are cruelly injured or killed by dogs and coyotes every year.

Quarantine incoming goats. Make no exceptions. House them at least 50 feet (15 m) from other goats but where the quarantined goats can see other animals (preferably goats). Quarantine for a minimum of 30 days, then disinfect the quarantine pen before you use it again.

Set up a hospital area for sick goats; don't let them stay with your other animals.

Be with does when they kid. Assemble a kidding kit and know how to use it.

Discuss vaccination protocols with your vet. At minimum, vaccinate using CD/T toxoid and give annual boosters to safeguard your goats against enterotoxemia and tetanus.

Have goats tested for CAE and CL (see pages 156–157) before adding them to your herd. And don't buy goats that have hoof rot.

Weighing Your Goat

To properly administer some pharmaceuticals and almost all dewormers, you need to know how much your goat weighs.

The most accurate way to determine his weight is to weigh him on a scale. If he's small, pick him up and weigh the two of you on a bathroom scale, then put him down, weigh yourself again, and subtract the difference. So easy! If your goat is too large to do this, call a vet. Many vets, even small-animal vets, have a walk-on scale at their practices; call and see if you can weigh your goat there.

The alternative is to tape-weigh your goat. To do so, restrain your goat, place a dressmaker's tape around his body just behind his front legs, and pull it up until it's snug but not tight. Use the following chart to find the weight that corresponds with your measurement (the chart is calibrated for dairy-type goats; meat goats will weigh somewhat more). This chart has been around forever, and it works!

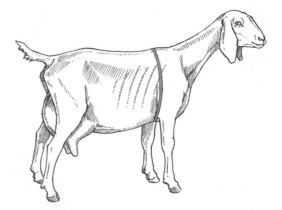

To tape-weigh a goat, pull a tape measure snug around his body as shown and refer to the table at right.

GOAT WEIGHT ESTIMATES

Girth/inches	lbs (kg)	Girth/inches	lbs (kg)	Girth/inches	lbs (kg)
10¼	4.25 (1.9)	22¾	42 (19)	35¼	130 (59)
10¾	5 (2.25)	23¼	43 (19.5)	35¾	135 (61.2)
11¼	5.5 (2.5)	23¾	48 (21.8)	36¼	140 (63.5)
11¾	6 (2.75)	24¼	51 (23.1)	36¾	145 (65.8)
12¼	6.5 (3)	24¾	54 (24.5)	37¼	150 (68)
12¾	7 (3.2)	25¼	57 (25.9)	37¾	155 (70.3)
13¼	8 (3.6)	25¾	60 (27.2)	38¼	160 (72.6)
13¾	9 (4.1)	26¼	63 (28.6)	38¾	165 (74.8)
14¼	10 (4.5)	26¾	66 (29.9)	39¼	170 (77.1)
14¾	11 (5)	27¼	69 (31.3)	39¾	175 (79.4)
15¼	12 (5.4)	27¾	72 (32.7)	40¼	180 (81.7)
15¾	13 (5.9)	28¼	75 (34)	40¾	185 (83.9)
16¼	15 (6.8)	28¾	78 (35.4)	41¼	190 (86.2)
16¾	17 (7.7)	29¼	81 (36.7)	41¾	195 (88.5)
17¼	19 (8.6)	29¾	84 (38.1)	42¼	200 (90.7)
17¾	21 (9.5)	30¼	87 (39.5)	42¾	205 (93)
18¼	23 (10.4)	30¾	90 (40.8)	43¼	210 (95.3)
18¾	25 (11.3)	31¼	93 (42.2)	43¾	215 (97.5)
19¼	27 (12.25)	31¾	97 (44)	44¼	220 (99.8)
19¾	29 (13.2)	32¼	101 (45.8)	44¾	225 (102.1)
20¼	31 (14.1)	32¾	105 (47.6)	45¼	230 (104.3)
20¾	33 (15)	33¼	110 (49.9)	45¾	235 (106.6)
21¼	35 (15.9)	33¾	115 (52.2)	46¼	240 (108.9)
21¾	37 (16.8)	34¼	120 (54.4)	46¾	245 (111.1)
22¼	39 (17.7)	34¾	125 (56.7)	47¼	250 (113.4)

Dealing with Internal Parasites

Goats are troubled by four types of internal parasites: stomach worms, lungworms, tapeworms, and liver flukes, and occasionally meningeal worm. Only roundworms (a type of stomach worm) are common; all goats have them to some degree.

Stomach worms. Depending on their species, stomach worms live in the stomach or small intestine, where they feed on blood and body fluids and interfere with the digestion and absorption of feed. Barber pole worm is common and potentially deadly. Badly infested goats are thin and they usually cough. They're weak and often anemic, and most of them suffer from diarrhea. Because they're run-down, they're open to infectious disease. Kids are more susceptible to roundworms than are adults.

Lungworms. Lungworms cause problems for goats who graze on boggy pastures inhabited by snails or slugs; these are the lungworms' intermediary hosts. Affected goats sometimes cough and wheeze. Severe infestations cause fluid in the lungs.

Tapeworms. Even substantial numbers of tapeworms don't cause adult goats much harm, but they can drastically affect the growth of kids. Heavy tapeworm infestations, however, occasionally clog or block an adult goat's intestinal tract.

Liver flukes. Like lungworms, liver flukes favor wet, low pastures. Their intermediary hosts are also snails. Low to moderate infestations of liver flukes impact growing kids; heavy infestations can kill adult goats.

Meningeal (brain) worms. These worms affect all types of ruminants, wild and domestic. This parasite's natural host is the white-tailed deer but its larval intermediaries are snails and slugs. Meningeal worm larvae travel up an infected goat's spinal nerves to his spinal cord and brain. This damages the central nervous system, eventually causing death.

Internal parasitism is a complex topic; it's best to discuss the subject with your vet. When you do, take along some fecal (manure) samples from your goats. The vet can run fecal egg counts, assess your needs, and devise a deworming schedule tailored for your herd.

Trimming Hooves

If your goat's hooves get long and roll under, he may limp. When his feet hurt, he won't browse as much or as often as his peers, so he'll also lose weight.

Trimming your goat's hooves is the most important thing you can do to keep them healthy. Soil moisture and type, time of year, and breed type influence how fast hooves grow, but as a rule of thumb, plan to provide a pedicure at least three or four times a year. It's a good idea to time pedicures to coincide with other labor-intensive procedures such as deworming and vaccinating. However, avoid trimming hooves during high-stress periods such as shearing (for fiber goats), extreme weather conditions, late pregnancy, and weaning. Hooves trim more easily when

---

Coccidiosis: Parasite or Disease?

Coccidiosis is a potentially fatal, highly contagious disease of goats, particularly young kids (goats develop a certain amount of immunity as they age). Other livestock species are afflicted by coccidiosis, but the protozoans that cause it are species-specific; your goat can't catch chicken or dog coccidiosis and vice versa. A single infected goat can shed thousands of microscopic oocysts (egg-like structures) in its droppings every day. If another goat ingests a sporulated (mature) oocyst, he can become ill a week or two later. As oocysts multiply and parasitize the gut, they destroy their new host's intestinal lining. Among the symptoms are watery diarrhea (scouring), sometimes containing blood or mucus; listlessness; poor appetite; and abdominal pain.

Dewormers don't kill coccidia; sulfa drugs are usually the treatment of choice. Talk to your vet about prevention and treatment.

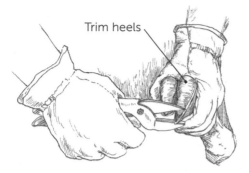

Step 2. Trim excess hoof growth here.

Step 4. A finished hoof should be perfectly level.

they're moist, whether from dew, rain, and snowmelt.

You'll need the proper tools to do the job easily and correctly. Most folks use hoof trimmers; others swear by horse-hoof nippers, a hoof knife, and a rasp. Shears or a hoof knife followed by a hoof plane work well, too. Getting it right is a matter of taste, experience, and convenience; when possible, use familiar tools that you have on hand. If you're buying tools, hoof shears are the best option: they're inexpensive and easy to handle.

1. Restrain your goat. Then you can squat beside him, perch on an overturned bucket, or stand and lean over to trim.

2. Begin trimming at the heel and work forward. Trim ragged edges even with the frog (the soft, central portion of each toe), then trim the walls level to match.*

3. If the frog is especially ragged, you can touch it up with a knife, taking paper-thin slices until you reach a hint of pink (it's a sensitive structure, so don't cut too far).

4. When you're finished, the hoof should be flat on the bottom and parallel or nearly parallel to the coronary band.

*When trimming a goat with foot disease, trim the bad hoof or hooves last. Otherwise, you'll spread disease to his healthy hooves. When you're finished, disinfect your tools to avoid infecting the rest of your herd.

The Perfect Hoof

To know how to trim a goat's hooves, examine the hooves of a 2- to 3-week-old kid from the bottoms and the sides. Trim at those angles and proportions; that's an ideal hoof.

Preventing
and Treating Hoof Rot

Hoof rot (sometimes called foot rot) is caused by an interaction of two bacteria, *Bacteroides nodosus* and *Fusobacterium necrophorum.* One of the bacteria, *F. necrophorum,* is commonly present in manure and soil wherever goats, sheep, or cattle are kept; it's only when *F. necrophorum* forms a synergistic partnership with *B. nodosus* that hoof rot occurs. *Fusobacterium necrophorum* can live in soil for only 2 to 3 weeks; it can, however, live in an infected hoof for many months.

Goats with hoof rot are very lame. They may hold up an infected foot and hop on three legs; if one or both forefeet are infected, they kneel to feed. Trimming infected hooves exposes a putrid-smelling, pasty goo lodged between the horny outer surface of the hoof and its softer inner tissues. You'll recognize hoof rot by its stench.

Hoof rot is spread from infected hooves to soil to healthy goats and sheep, so isolate all infected animals. To expose the disease-causing bacteria to oxygen, trim back hooves to infected areas and remove as much "rot" as you can. Treat according to your veterinarian's directions.

• • • • • • • • • • • • •

Judge not the goat by his horns.

~ Chinese proverb

• • • • • • • • • • • • •

Don't buy goats or sheep from sale barns or from infected herds. Make certain any commercial transporters who carry your livestock thoroughly disinfect their trailers after every trip. Disinfect your shoes after visiting other facilities. Keep barnyards and holding areas as dry as you can.

The Major Diseases

Most goats are healthy goats; however, you'll want to know about the three major diseases that affect them: caprine arthritic encephalitis (CAE), caseous lymphadenitis (CL), and Johne's disease. Make sure that animals have been tested for these diseases before you purchase them, and know how to identify the symptoms should one of your goats somehow become infected.

Caprine Arthritic Encephalitis (CAE)

CAE is a goat-specific disease related to ovine progressive pneumonia (OPP) in sheep. First identified in 1974 and initially called viral leukoencephalomyelitis of goats (VLG), it's a progressively crippling disease caused by a retrovirus. Unfortunately, there is no vaccine against CAE and no cure. There are tests, however, to single out infected animals. It's always best to buy from annually tested, CAE-free herds or have goats you plan to purchase tested at your own expense.

The virus that causes CAE is responsible for a neurological disease of the spinal cord and brain in young kids and a joint affliction of older goats. Kids affected with CAE show signs of disease between 1 and 4 months of age. The virus causes progressive weakness of the hind legs, leading to eventual paralysis in a few days or up to several weeks. Kids remain in good spirits and continue to eat and drink.

The arthritic form of CAE usually surfaces between 1 and 2 years of age. Some affected goats are badly crippled in a few months; in others, the disease progresses slowly over a span of years.

The virus that causes CAE is transmitted to a kid from its infected dam through her colostrum and milk. To prevent transmission of the virus, offspring of CAE-positive dams must be taken away at birth, before they suckle, and bottle-raised on properly heat-treated colostrum and then milk from a CAE-free goat, a ewe, or a cow, or, if colostrum is not available, on an IgG supplement, followed by CAE-free milk or milk replacer. (IgG belongs to a class of immunoglobulins.)

Caseous Lymphadenitis (CL)

Caseous lymphadenitis (called cheesy gland in Britain and elsewhere) is caused by the bacterium *Corynebacterium pseudotuberculosis*. It's mainly a problem of sheep and goats, and the two species can easily infect one another. CL manifests as thick-walled abscesses packed with odorless, greenish-white, cheesy-textured pus. CL abscesses form on lymph nodes, particularly on the neck, chest, and flanks, but also internally on the spinal cord and in the lungs, liver, abdominal cavity, kidneys, spleen, and brain. CL is contagious and incurable. Transmission is through pus from ruptured abscesses.

However, don't panic if your goat develops a lump. Few lumps are caseous lymphadeni-

tis. Goats are notorious for developing CL-like nodules on injection sites. Abscesses can also form when any of hundreds of organisms breach the skin by way of puncture wounds, splinters, and cuts or abrasions. The only way to be sure an abscess is CL is to have its contents cultured. Quarantine any goat with a progressively softer-centered lump, lance the abscess, and drain some pus into a sterile container to take to your vet. Wear gloves and don't let pus smear the surroundings or drip on the ground. The goat should remain in quarantine until the abscess has healed.

Buy only from CL-free herds and flocks. CL-positive sheep and goats can be vaccinated with autogenous vaccine custom-made using pus from one of your infected animals; this will suppress future abscesses. Colorado Serum's Case-Bac is an effective vaccine for healthy sheep, but there is no species-specific vaccine for goats.

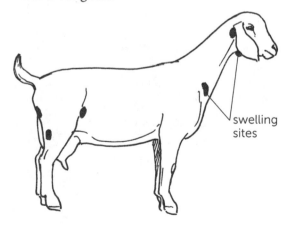

CL usually occurs on lymph nodes. Suspect abscesses that form at these locations.

"PADDY MCGINTY'S GOAT"

Perhaps not as famous as "Bill Grogan's Goat," "Paddy McGinty's Goat" is just as much fun. A traditional folksong, it has at least 20 stanzas. These are just a few.

Mister Patrick McGinty
An Irishman of note,
Came into a fortune and
Bought himself a goat.
Said he, "Sure of goat milk,
I aim to have my fill."
But when he got his nanny home,
He found it was a bill.

• • • •

Now Pat McGinty's goat
Had a wondrous appetite,
And often for breakfast
He'd eat some dynamite.
A box full of sparklers
He'd swallow with a grin
And when he'd had his fill of that
He drank some paraffin.

• • • •

He sat down by the fireside,
He didn't give a hang.
He swallowed a spark
And exploded with a bang.
So when you get to heaven
You can bet your dollar note
That the angel with the whiskers on
Is Pat McGinty's goat.

Johne's Disease

Johne's (pronounced *YO-nees*), also called paratuberculosis, is a contagious, slow-developing, progressively fatal disease, of cattle, sheep, goats, deer, antelope, and bison; it's most commonly seen among dairy cattle. Among the symptoms are progressive weight loss and weakness that leads to death.

Johne's is caused by *Mycobacterium paratuberculosis,* a close relative of a bacterium that causes tuberculosis in humans, cattle, and birds. According to Johne's Information Center statistics, 7.8 percent of beef herds and 22 percent of dairy herds in the United States are infected with *M. paratuberculosis.* Johne's disease typically enters goat herds when bottle kids are fed milk from infected cattle or an apparently healthy adult carrier is added to the herd. The infection then spreads to other goats, generally when young kids nibble contaminated manure from an infected goat.

An effective test is readily available, although few goat breeders test for Johne's. If a goat you're interested in buying hasn't been tested, consider having her tested before you buy.

Finding a Veterinarian

Don't wait for an emergency to choose a veterinarian. Finding a knowledgeable vet who will treat goats is probably harder than you think. Small-animal vets tend to consider them livestock; equine practitioners and large-animal vets don't see enough goats to know how to treat them well.

The best way to find a goat-savvy veterinarian (and eliminate the ones you probably shouldn't use) is to pick the minds of other goat owners: Why do they use the vet they use? Have they used other vets? Why did they switch? This is constructive gossip, so stick with it and learn all you can. If you don't know anyone to ask, haul out your phone book and start calling veterinarians.

Ask prospective veterinarians if they are interested in goats and have practical experience treating them. If they don't, ask about their expertise with similar small ruminants such as sheep, alpacas, and llamas. If they have none, ask if they'd be willing to learn (this is important). Would they be comfortable with input from other vets (many experts consult over the phone)? And would they object if you research goat-related topics and bring applicable resources to their attention?

Also ask whether they make farm calls or if you'd have to haul your goats to their practice. Will they carry a tab or do you pay for routine services as they're rendered? What about emergencies? Will they set up a payment schedule if you can't cover the tab all at once?

Once you find a vet you like, schedule a routine farm call or an office visit to get a feel for the way the vet and staff handle your goat. If you may have to board your goat at the vet's practice, ask to see the facilities. Can you bring your goat his own feed from home? Could he see (but not interact with) other animals, so he wouldn't be stressed or lonely?

You will be paying the bill. If any vet is uninterested in your goat or seems incompetent, or if you can't quite stand the vet's "stall-side manner," keep looking.

Once you find a good vet, keep him. Listen carefully, follow treatment protocols, and pay your bill on time. If something he does upsets you, tell him; don't complain to everyone else you know. Treat him or her well. Good goat vets are gold-plated; anyone who loves goats will tell you so.

Being Your Own Vet (Sometimes)

When your goat is sick, take him to a vet. The symptoms of many serious goat diseases resemble one another to an alarming degree, such as pregnancy toxemia and milk fever, listeriosis and goat polio, tetanus and rabies, and enterotoxemia and bloat. Treating for one disease when your goat has another won't help your goat.

Many pharmaceuticals used to treat goats are prescription drugs, some of them off-label for goats. Most vets won't hand them out without seeing your animals. Some dispense prescription drugs to established clients, but the bottom line is that you'll have to get most drugs through a vet.

That said, many vets appreciate clients who can do the basic procedures themselves, such as vaccinating and deworming, treating wounds, and administering antibiotics on the vet's orders.

If you live where you can't find a suitable vet, you'll have to learn to do more-serious treatments yourself. This is not without peril, but it can be done. Buy a few good books about goat medicine, but also visit the websites listed at the back of this book. Print out information you may need and file it so you can find it in a hurry. And join goat-related e-mail lists (see Resources) where you can ask advice in an emergency. Another solution: Find a mentor to help when you need it, preferably someone who lives nearby and can show you the ropes firsthand.

Checking Vital Signs

When your goat is sick and you call for help, be ready to provide your goat's vital signs. These are normal values for adult goats; kids' values run slightly higher.

Temperature: 101.5 to 104.5°F (38.5–40.3°C)
Heart rate: 70 to 90 beats per minute
Respiration rate: 12 to 20 breaths per minute

Taking a Goat's Temperature

An elevated temperature usually indicates infection or dehydration, though a goat's temperature rises slightly as the day progresses and may be higher by a full degree on hot, sultry days. A subnormal temperature can mean hypothermia, hypocalcemia (milk fever), or that the goat is dying.

You need a digital rectal thermometer to take a goat's temperature. Veterinary models are best, but a thermometer designed for humans works, too.

1. Restrain the goat: Tie him using his collar and lead, push him against a wall, and hold him there, or use a milking or grooming stand. Find someone to hold him still. It's easiest to place a small kid facedown across your lap.

2. Insert the business end of a lubricated thermometer (KY Jelly, Vaseline, birthing lubricants, and mineral or vegetable oils are excellent lubes; your own saliva will work in a pinch) about 2 inches (5 cm) into a standard-size adult goat's rectum (correspondingly less for kids and minis; length will vary by breed). Don't grasp the goat's tail; he won't like it.

3. Hold the thermometer in place until it beeps (or wait 2 to 3 minutes if using an old-fashioned, nondigital thermometer).

4. When you're finished, swab the thermometer with an alcohol wipe and return it to its case. Store it at room temperature.

Nannyberries

GOAT HAIKU

Goatling observes sheep
You guys look like fuzzy rugs
Better to be goat

• • •

We descend from ibex
No more mountains to conquer
Now we climb on cars

• • •

Does ruminating
Wisdom of the ancient ones
Shines in amber eyes

• • •

Soft sweet hay and pellets
Warm straw bedding in wintertime
A goat's life is gooooood

A WELL-EQUIPPED FIRST-AID KIT

Every goat owner needs a first-aid kit. Pack it in a container that's easy to carry (we use a 5-gallon food-service bucket with lid, but a plastic storage or fishing tackle box, a backpack, and the like work well, too) and stow it where you can easily find it. A basic kit might contain:

Bandaging supplies. Several rolls of self-stick disposable bandage such as Vet-Wrap, Telfa pads, gauze sponges, a roll or two of 2½-inch (6.4 cm) sterile gauze bandage, 1-inch and 2-inch (2.5 cm and 5 cm) rolls of paper adhesive tape, a partial roll of duct tape, and several high-absorbency sanitary napkins — they're great for applying pressure bandages to staunch bleeding.

Blood Stop powder. Flour or cobwebs also work well.

Cleaning solutions. Saline solution and alcohol for sure; packaged alcohol wipes and Betadine Scrub are also useful.

Disposable syringes and needles. Syringes in 3 cc, 10 cc, and 60 cc sizes; 16-, 18-, and 20-gauge needles

Epinephrine. Always keep it on hand; also by prescription from your vet.

Flashlight. One, or better, two

Hardware. A digital thermometer, blunt-tipped bandage scissors, regular scissors, a hemostat, a stethoscope, a hoof pick

In addition to your basic first-aid kit you'll probably want to keep some or all of these items in your refrigerator or a medicine cabinet in the barn.

Banamine (a prescription pain and inflammation reliever from your vet). If you don't have any, baby aspirin is a distant second-best for pain. Dosage is the same pound for pound as given for a human infant.

Injectable antibiotics. Several types; ask your vet which to keep on hand.

Injectable thiamine. Vitamin B_1, an important nutritional booster when goats are sick. It's a prescription item, so get it from your vet.

Pepto-Bismol. Used mainly for diarrhea in nursing kids. Give 2 cc every 4 to 6 hours for newborns and 5 cc every 4 to 6 hours for older kids.

Probiotic gel or paste. Choose a ruminant-specific product like Pro-bios or FastTrack.

Spare collars and a lead

Wound treatments. Some good ones are Schreiner's Herbal Solution, emu or tea tree oil, Betadine, Neosporin, and topical eye ointment (for injured eyes).

Checking Heart Rate

The easiest way to check a goat's heart rate is with a stethoscope. If you don't have one, take his pulse by pressing two fingers against the artery just below and slightly to the inside of the edge of the jaw, two-thirds of the way back from the muzzle, or against the large artery on the inside of either rear leg up near the groin. Count the number of pulses in 15 seconds and multiply by 4.

Assessing Respiration

Watch your goat's rib cage. Count the number of breaths he takes in 15 seconds and multiply by 4. Extreme heat and fear or anger elevate pulse and respiration. Slightly elevated readings are sometimes the norm.

Giving Shots

Learn to vaccinate your goats and to give shots. It sounds hard but is easier than you might think. Ask your veterinarian or an experienced goat or sheep breeder to show you which pharmaceuticals to use and how and where to give injections.

Assembling the Right Supplies

Buy needles, syringes, and vaccines from your vet, at feed stores and farm stores, or by mail order from farm supply warehouses. Some states require all vaccines to be prescribed and filled by a vet, and some products (such as rabies vaccine in New York) must even be administered by a vet. Check your state laws.

Select the correct pharmaceutical and read or reread the label. If a product was inadvertently stored incorrectly, discard it. Also check the expiration date; don't use a product if it's outdated.

Choose the right disposable needle for the job. Almost all goat injections are given subcutaneously (SQ; under the skin) and should be given using a 18- or 20-gauge needle that is ½ inch or ¾ inch long.

You'll need a new needle for each goat, plus a transfer needle to stick through the rubber cap on each product (for example, if you vaccinate three goats using CD/T toxoid and rabies vaccines, you'll need five needles). A shot is less painful for the goat if you use a new needle each time, and it eliminates the possibility of transmitting disease (such as CAE) by way of a contaminated needle.

Nannyberries

"GOAT" IN DIFFERENT LANGUAGES

Chinese: yáng

Czech: kosa (male goat — kozel)

Filipino: kambing

French: chèvre (male goat — bouc)

German: Zienge (male goat — Ziegenbock; kid goat — zicklein)

Hindi: bakara (male goat — bakarl)

Irish Gaelic: gabhar

Italian: capra (male goat – caprone or capro; kid goat — capretto)

Japanese: yagi

Russian: koza

Spanish: cabra (male goat — macho cabrío or chiva; kid goat — cabrito)

Urdu: bakri

Preparing the Needle

Use the smallest disposable syringe that will do the job; it will be easier to handle than a big, bulky syringe, especially for women with small hands. Never try to sterilize disposable syringes; boiling compromises their integrity.

1. Jab a new, sterile transfer needle through the cap of each pharmaceutical bottle. *Never* poke a used needle through the cap to draw vaccine or drugs!

2. Attach your syringe to the transfer needle. When you're going to draw fluid from the bottle, first inject air into the bottle — half the amount of the prescribed dose (e.g., if you're drawing 8 cc of fluid, inject 4 cc of air) — to avoid the considerable hassle of drawing fluid from a vacuum; then pull a tiny bit more of your required fluid amount into the syringe.

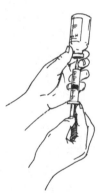

3. Detach the syringe from the needle in the bottle and attach the needle you'll use to inject the goat.

4. Holding the syringe with the needle end pointing up and away from you, press out the small amount of excess fluid to remove any bubbles that were created as you drew the fluid out of the bottle.

Preparing the Injection Site

Choose the best injection site. You want to inject the vaccine or drug where it will work well but not injure your goat. Preferred sites for subcutaneous injections are the neck, over the ribs, and especially into the loose skin in the "armpit" area. When injecting a large volume of fluid (generally 10 cc or more for an adult goat), break the dose into smaller increments and inject it into more than one site.

Swab the injection site with alcohol (prepackaged alcohol swabs are easy to use). Never inject into damp, mud- or manure-encrusted skin. Always make sure your goat is fully restrained before giving a shot.

Giving a Subcutaneous (SQ) Injection

When giving an injection, it's okay to use the same needle and syringe to inject the next vaccine or drug into the same goat, but always use a new transfer needle.

1. Pinch up a tent of skin and poke the needle into it, parallel to your goat's body — never at a right angle. Take care not to shove the needle through the tented skin and out the opposite side. Also be very careful not to prick the muscle mass below it.

2. Slowly depress the plunger and when the syringe is empty, withdraw the needle.

3. Rub the injection site to help distribute the drug or vaccine.

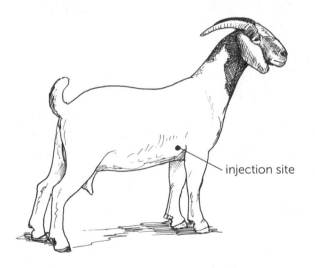

injection site

An ideal subcutaneous injection site is into the loose skin of the armpit area. Injecting into a fold of skin on the neck or over the ribs works too.

Goats often develop benign injection-site nodules after receiving a shot. Remember where you gave shots, so that you don't panic if one pops up. Your goat's body usually reabsorbs these lumps over time, but occasionally an abscess forms. If there is any chance the lump could be CL, treat it accordingly (see page 157). Otherwise, lance and drain it and treat it as a wound.

Don't Forget the Epinephrine

Epinephrine, also called adrenaline or epi, is a naturally occurring hormone and neurotransmitter manufactured by the adrenal glands. It's used to counteract the effects of anaphylactic shock, a serious and rapid allergic reaction that can kill.

Any time you give an injection, no matter the product or amount injected, you should be prepared to administer epinephrine to counteract an anaphylactic reaction. If a goat goes into anaphylaxis (symptoms include glassy eyes, increased salivation, sudden-onset labored breathing, disorientation, trembling, staggering, and collapse), you won't have time to race to the house to grab the epi. You might not even have time to fill a syringe. You have to be ready to inject it *right then*.

If you give your own injections, get a bottle of epinephrine from your vet (it's a prescription item) and keep a dose drawn up in a syringe in the refrigerator. Kept in an airtight container (we use a clean glass jar with a tight-fitting lid), a syringe of epi keeps as long as the expiration date on the epinephrine bottle. Take the loaded syringe with you every time you give a shot. Standard dosage is 1 cc per 100 pounds; don't overdose, as it causes the heart to race.

Basic Procedures for Kids

You needn't be an expert to raise kids, but there are a few procedures you'll want to learn (or at least learn about) to keep your caprine friends in the pink. Here we talk about bottle-feeding, disbudding, and castrating kids.

Bottle-Feeding Baby

If the kid is less than 24 hours old, make sure he has ingested enough colostrum. If you aren't sure, more won't hurt him. Heat colostrum by immersing the bottle (or freezer baggie) in warm but not sizzling-hot water; never microwave the precious stuff. If you're fortunate, you'll buy or be given a bottle-trained kid, but you'll usually have to train your kid yourself. The best (and easiest) way is the following method we use. If your kid can't suck, you'll need to tube-feed him (see page 189).

1. Sit cross-legged on the floor with the kid tucked between your legs and facing away from you, his front legs straight and his butt on the ground.

2. Cup your left hand under his jaw and open his mouth. Insert the nipple with your right hand, then balance and steady it with the fingers of your left hand, with the left palm still under his jaw. This way he's less likely to spit out or otherwise lose the nipple.

3. Hold the bottle at mama-goat height (this height will vary by breed). Encourage him to lift and tip back his head. Kids function as single-stomached animals rather than ruminants until they begin nibbling dry food, usually around 2 to 3 weeks of age.

Starting a bottle kid is easiest if you sit on the floor with him propped between your legs, facing outward. Use a nipple that he likes. We prefer the Pritchard nipple, at the center of the lineup above. You could also use a soft, pliable one designed to be pulled over the mouth of a pop bottle (left), or a nipple designed for human infants (right).

When your kid's head is up and back, a band of muscle tissue closes and, bypassing his undeveloped rumen, directs milk straight from his mouth to his abomasum (the only stomach compartment containing digestive enzymes). However, don't let milk pour into his mouth; if he can't swallow the milk fast enough, it might spill into his lungs. Hold the bottle as level as possible while still keeping fluid in the bottle cap and nipple.

4. If your kid won't nurse, place a towel over his head while feeding (this simulates the darkness of a doe's underbelly). Or squeeze a little fluid from the nipple and use it to dab milk on his lips. Give him a taste but don't shoot a stream into his mouth.

An older kid who has already nursed from his mom is notoriously hard to bottle-train. Persevere. If he still refuses to nurse and he's an otherwise healthy young goat, place the kid back in his quarters and wait an hour or two before trying again.

Whether bottle or tube feeding, measure the amounts you offer your kid and feed uniform meals at evenly spaced intervals. Kids require 15 to 20 percent of their body weight in milk or formula daily. Be sure to wash the bottle and nipple thoroughly after each feeding. This schedule works for an average dairy goat kid.

Days 1 to 2. Feed 2 to 3 ounces of warm (102°F [39°C]) colostrum, or formula with colostrum replacer added, per feeding, six times in 24 hours.

Days 3 to 4. Feed 3 to 5 ounces of warm milk or formula per feeding, six times in 24 hours; also, place a small, spill-resistant container of drinking water in his quarters and change it frequently to keep it fresh.

Days 5 to 14. Feed 4 to 6 ounces of warm milk or formula, four times in 24 hours, gradually switching to milk or formula straight from the refrigerator.

Days 15 to 21. Feed 6 to 8 ounces of milk or formula, four times in 24 hours; introduce green pasture or free-choice leafy alfalfa or high-quality grass hay and small amounts of high-protein (16–20%) dry feed.

Days 22 to 35. Gradually work up to 16 ounces of milk or formula fed three times in 24 hours; continue until the kid is 6 weeks old.

Weeks 6 to 8. Make certain the kid is eating sufficient amounts of forage (grass or hay) and chewing his cud. At 6 weeks, begin decreasing the amount of liquid offered at two feedings, eliminating these two feedings altogether by the end of week 8; leave the remaining feeding at 16 ounces.

Weeks 8 to 12. Continue feeding 16 ounces until the end of week 11; in the next seven days, gradually eliminate the last feeding of the day.

Never forget the cardinal rule of bottle feeding: it's easy to kill your kid with kindness. If your kid cries between meals (and he probably will), you'll be tempted to feed him snacks or bigger portions. Don't do it. A kid can digest only so much milk or formula. If you think he needs between-meal snacks, bottle-feed him Pedialyte, Gatorade, a simple livestock electrolyte supplement such as ReSorb, or plain old water, all of which help keep him hydrated and satisfied but don't require lengthy digestion times.

Disbudding

Most dairy goat breeders disbud their kids; meat goat producers and packgoat breeders generally don't. Disbudding consists of destroying a kid's emerging horn buds by burning them with a red-hot disbudding iron to prevent the horns from growing. It is done as soon as a hard bump the size of a pimple can be felt beneath the kid's skin; this may be at 3 days to as much as a week and a half of age, depending on sex (bucklings' horn buds mature faster than doelings') and breed.

Only experienced operators should attempt disbudding: if the iron is held in place too long, it will damage the kid's brain; not long enough, and scurs result. Scurs are ugly, misshapen horn remnants that usually require trimming and sometimes curl around and grow into a goat's head. If you don't want your kid to have horns, make arrangements for a goat-savvy veterinarian or experienced goat person to do the procedure when the kid is a few days to a week or so old. And never attempt to dehorn an adult goat. Doing so will cause massive bleeding and gaping holes that won't heal for months.

Castrating

Unless you breed goats and accept the considerable hassle of keeping your own buck, you'll probably want to have your male kid castrated. Wethers make the best pets and recreational goats. They're easier-going than does and bucks, their moods don't fluctuate, and, unlike uncastrated males, they won't reek or get feisty when breeding season rolls around.

Many goat producers "band" male kids to castrate them. This consists of applying a small, strong, doughnut-shaped rubber band around the top of a kid's scrotum using an inexpensive applicator tool called an elastrator. The band prevents circulation to the scrotum, causing it to atrophy and slough off in 4 to 6 weeks. A plus: Banding costs pennies and anyone can do it. A huge minus: Regular-size elastrator bands must be applied when kids are very young, between 1 and 4 weeks of age, making banded kids more likely to develop urinary calculi (see page 149) as adults. Proponents of banding claim it's painful for only 5 or 10 minutes (until the area becomes numb), whereas opponents insist it causes days of stress and pain. It's possible to band an older kid using a larger banding device designed for calves, but an older kid should be given an injection of a prescription drug called flunixin meglumine (Banamine) prior to banding to ease initial pain. Baby aspirin (dosed according to body weight) dissolved in the youngster's next few bottles helps, too.

You could also wait until your kid is 4 to 6 months old and have your veterinarian surgically castrate him using an anesthetic. Goats don't fare well under general anesthesia, however, and a local anesthetic doesn't completely ease the pain of surgical castration. We opted for this procedure only once, then went back to banding with the larger bander and Banamine.

Common Kid Problems

Healthy baby goats are bright and alert. They're charged with energy and their coats show a healthy sheen. If your kid is lethargic and his coat seems dull, if discharges of any sort glob his eyes or nose, or if he hangs his head and his tail droops instead of curling over his back, something is definitely wrong. Don't experiment with home cures. When your goat is sick, take him to a vet.

That said, anyone raising a bottle kid should learn to take his temperature and to recognize the early symptoms of rapidly progressing kid-killers such as floppy kid syndrome, enterotoxemia, pneumonia, and scours.

Itinerant preacher Charles (Ches) McCartney, better known as The Goat Man, traveled the United States from 1930 to 1968 in a ramshackle wagon pulled by goats. McCartney supported himself and his son, Albert Gene, shown with McCartney's first wife in this early photo, in part by selling postcards and allowing photographers to shoot photos of his entourage.

Floppy Kid Syndrome and Enterotoxemia

Authorities believe that floppy kid syndrome (FKS), first recognized in the late 1980s, is caused by a type of clostridial bacterium similar to the type that causes enterotoxemia and tetanus. Such organisms are normally present in the gastrointestinal tracts of goats, but under certain conditions they can go berserk and multiply rapidly while producing a load of lethal toxins.

Floppy kid syndrome and enterotoxemia can occur when a bottle kid ingests too much milk for his system to handle. Milk pools in his stomach and curdles, leading to acidosis, indigestion, and slowing of intestinal peristalsis. Toxins build in the gut, and unless aggressively treated, the kid will die. Floppy kid syndrome affects young kids between 3 and 21 days of age. Early symptoms are a distended belly, stumbling, wobbly gait, and extreme lethargy. At the height of the disease, afflicted kids are totally limp.

Enterotoxemia is generally accompanied by watery, evil-smelling diarrhea; high fever; and severe abdominal pain. If your kid's dam was given a routine CD/T booster (to protect her from enterotoxemia types C and D and tetanus) a few weeks prior to kidding, he'll have gained some immunity through ingesting her colostrum. He should, nevertheless, still be vaccinated with CD/T toxoid at 7 or 8 weeks of age and again 4 weeks later.

If your kid displays symptoms of floppy kid syndrome or enterotoxemia, stop feeding him milk or formula, substitute livestock electrolyte, Pedialyte, or Gatorade, and get him to a vet ASAP.

Pneumonia

Pneumonia is caused by several strains of bacteria and viruses. It can be brought about by exposure to sick goats; being housed in a cold, wet, poorly ventilated, dusty, or drafty environment; stress of any sort; excessive worm loads; or drastic weather changes. Milk dribbling into a kid's lungs can also trigger pneumonia.

A kid with pneumonia will generally cough and run a fever (anything above 103.5°F [39.5°C] should be suspect); his breathing may be rapid and shallow, raspy and labored, or, rarely, totally unaffected. Any nasal discharge will be thick and greenish yellow rather than clear or white. He probably won't eat. He might stand or lie hunched over with his hair coat ruffled, his tail at half-mast, and his head hanging down. If you suspect pneumonia, give the kid half a baby aspirin to help bring down the fever and rush him to a vet.

Scours

Scours is another name for diarrhea. It's usually a symptom of a larger, more serious problem (enterotoxemia, coccidiosis, pasteurella, salmonella, *E. coli,* and severe worm infestation, to name a few) but can be caused by simple mismanagement, too. In the latter case, the most frequent causes are overfeeding, using low-quality milk replacers or dirty feeding utensils, and sudden changes in feed or feeding schedules. If diarrhea is watery, foul-smelling, and won't let up, or your kid's temperature is above or below the norm, rush him to a vet for proper diagnosis and treatment.

If scouring is mild and the kid's temperature is within normal parameters (101.5 to 103.5°F [38.5–39.5°C]), treat mismanagement-induced diarrhea by correcting its cause and feeding Pedialyte, Gatorade, or livestock electrolyte products such as ReSorb in lieu of milk for no more than a day or two. Remember, very young kids make sticky, yellow, pudding-consistency poo for a week or more. That's normal; it's not scours.

• • • • • • • • • • • • •

*Don't approach a goat
from the front, a horse from
the back, or a fool from
any side.*

~ Yiddish proverb

• • • • • • • • • • • • •

CHAPTER 11

You've Got to Be Kidding!

Every kid is a gazelle in the eyes of its mother.

~ Moorish proverb

Unless your doe milks through for the rest of her life (a very unusual circumstance indeed), she'll have to give birth to kids to produce milk. For most does this is an annual event. And it's fun! But you'd best be prepared. Here's what you need to know.

The (Caprine) Birds and Bees

From early autumn through midwinter, seasonal breeders (such as most dairy breed does) cycle, or come into heat, every 18 to 21 days; aseasonal breeders (such as Pygmies, Boers, and Nigerian Dwarfs) cycle year-round. Ovulation occurs 12 to 36 hours after the onset of standing heat. Signs of heat are:

- Interest in nearby bucks (and sometimes wethers)
- Loud and strident vocalizing
- Increased activity level (particularly fence walking)
- Tail wagging (also referred to as flagging)
- Frequent urination
- Mounting other does and wethers and being mounted by them
- Decreased appetite
- Lower milk production

Does are stimulated by the appearance and scent of a buck. A fully receptive doe stands with her head slightly lowered, her legs braced, and her tail to one side (this is called standing heat). She'll probably urinate if another goat sniffs her.

If you don't have a buck, collect a terry-cloth washcloth or a piece of terry-cloth towel and a container with a tight-fitting lid, then visit someone who has a buck in rut. Scrub the rag all over the buck, concentrating on urine-stained areas and the scent gland region

Nannyberries

FROZEN IN TIME

In 1871, Richard L. Maddox revolutionized photography by inventing the dry-plate process. Using this splendid new technology, traveling photographers no longer relied on cumbersome, portable darkrooms to ply their trade. And since the process was far more sensitive to light than the wet-plate process used up to that point, relatively fast shutter speeds were possible, making carefully held long poses a thing of the past.

During the early 1900s, affordable postcard-format cameras such as the Chicago Ferrotype Company's Mandel-ette postcard camera emerged. These were simple box cameras with fixed-focus lenses. In the back of a camera was a black changing bag through which the photographer moved an exposed paper negative to the built-in developer tank attached to the bottom of the camera. Pictures on demand! The best part was that in 1919, the Mandel-ette, complete with tripod and enough material for 116 postcards, cost the grand sum of $7.75 ($173.81 in today's money). Itinerant photographers snapped them up and took to the road.

Among them was a special class of men — children's photographers who traveled with a pretty pony, a donkey, or a handsome goat-drawn cart to act as props. Between the first decade of the new century and the early 1960s, thousands upon thousands of smiling youngsters posed for their portraits holding the reins of the photog-

A handsome young man poses with a black goat named Billy in this postcard-backed image captured by a traveling children's photographer.

rapher's sidekick — sometimes in costumes provided by the photographer and sometimes wearing the dress of the day. Since this was an era when few working-class families could afford a camera, picture-taking day, especially if they were being posed in a photographer's pretty goat cart, was a highlight of many young lives.

Antique picture postcards of children and a goat cart are popular collectibles today. Most are 3¼ inches × 5½ inches (8¼ cm × 14 cm) in size, but smaller postcard backs could be used in many cameras. One of their beauties is that they're inexpensive (few in my collection cost more than $15 apiece) and readily available. I find most of mine on eBay, but flea markets and antique shops are fertile hunting grounds, too. Older relatives often have these winsome photos tucked away. Ask!

of his forehead. When the washcloth reeks of a strong scent, pop it into the container and shut the lid. When you think your doe might be in heat, present her with the buck rag. Hold it to her nose and let her take some nice, deep sniffs. If she's in heat, she'll likely react with definite signs. It's that easy.

Once impregnated, she enters anestrus; most does stop coming into heat at this time. Between 144 and 157 days after the onset of her last standing heat (the precise time varies among breeds), she gives birth to one to five darling kids.

Choosing a Boyfriend for Your Doe

Breed your doe to a quality buck. If she's registered, her kids will be more valuable if you breed her to a registered buck of her own breed. To locate registered bucks available for breeding purposes, ask the person who sold you your doe or scope out directories at your registry's website (see Resources).

Choose your doe's mate as carefully as you would if you were buying him. Is his conformation correct for his breed and does it complement your doe's conformation? Does he have a good bite? Check his teats; you don't want his doeling daughters to inherit abnormalities from their dad. If he's a Boer or Savanna, he should have two or four teats; all other breeds should have two. They should be nicely shaped like a doe's teats except smaller, with no unusual bumps or deformities.

Before committing to a buck, visit the breeder's farm and look around. Do any of the breeder's goats have CL abscesses? Are they healthy, happy, and in good flesh? Ask about accommodations. Will your doe have her own pen or be shoved out to fend for herself with the rest of the breeder's herd? Are you comfortable with the response? If the buck lives nearby and you can tell when your doe is in standing heat, you can take her for a single breeding and bring her home the same day. Otherwise, she'll have to stay with the breeder until the deed is done.

Expect to pay a decent price to breed to a quality buck. Call the breeder when your doe comes in heat, deliver her promptly, and pick her up when she's bred. Then sit back and wait — the fun is just beginning!

Preparing for the Big Event

Feeding during late gestation is an art. We provide our goats with a high-quality 2:1 calcium-to-phosphorus mix, and we feed grain (in moderation) and alfalfa hay or pellets to late-gestation does. Here are other preparation tips.

To prevent pregnancy toxemia, does should be neither thin nor obese going into pregnancy. Don't overfeed in early pregnancy, but do increase carbohydrate (energy) intake during a doe's last trimester by adding judicious amounts of grain and/or alfalfa to her diet. Make sure does get enough exercise.

To prevent milk fever, provide pregnant does with 2:1 calcium-to-phosphorus mix

. . . .

Wherever a goat goes, her kid follows.

~ Mzab proverb

. . . .

PREGNANCY TOXEMIA AND MILK FEVER

Pregnancy toxemia is a potentially life-threatening metabolic disorder that afflicts does during their final weeks of pregnancy. It is caused by excess ketones (chemical substances that the body makes when it does not have enough insulin in the blood). These accumulate when a late-gestation doe ingests fewer calories than she needs to support herself and her fetuses, or when a milking doe produces more milk than her energy intake can provide for. Symptoms are a lack of appetite, depression, unwillingness to stand, muscle tremors, and staggering, as well as teeth grinding and other indications of pain. The condition eventually results in coma and then death.

Milk fever (hypocalcemia) is caused by a drop in blood calcium a few weeks prior to and immediately after kidding. Symptoms are similar to pregnancy toxemia and the two are easily confused. Both are life-threatening situations, so call your vet without delay. Treatment must be swift and aggressive.

We've never treated a case of milk fever on our farm, but pregnancy toxemia? Oh my! During our first kidding season, one Boer doe died — the only adult goat we've lost in 7 years. Another needed a C-section; a third was so weak by the day she went into labor that we had to pull her kids. Why did all this happen? Too much phosphorus and too little calcium in their diet made them sick. We changed their diet and then we didn't have any more problems with pregnancy toxemia.

at all times; feed alfalfa hay or pellets to late-gestation and lactating does.

Stop milking your doe at least 2 months before she's due to kid. Making milk is hard work. Her body needs a rest before she gives birth.

Feed her the regular ration until 6 weeks before delivery. Dramatic fetal growth doesn't occur until the final two months of pregnancy; feeding too much, too soon will make both your doe and her unborn kids too fat at birthing time. Fat, flabby does are prone to birthing problems and huge kids are hard to deliver.

Your doe needs good feed but not enough to make her obese.

Three to 5 weeks before she's due to kid, give your doe a booster shot of CD/T toxoid (a vaccine used to prevent enterotoxemia and tetanus in healthy goats). That way she's sure to pass antibodies to her kids through her colostrum. At the same time you boost your doe with CD/T toxoid, you may elect to inject her with a prescription selenium and vitamin E supplement called BoSe. This prevents white muscle disease in her kids and is a common practice in selenium-deficient parts of the country. Check with your vet or county Extension agent to see if it's needed where you live.

Plan to be with your doe when she kids. Though most deliveries go smoothly, those that don't can break your heart. Be there — no excuses.

Set up a cozy jug (mothering pen) where your doe can bond with her kids for 2 or 3 days. If you don't have a lactating doe to provide milk for unexpected bottle babies, use your favorite store-bought, milk-based bottle-baby formula or buy a bag of quality milk replacer to have on hand.

Assemble the Right Supplies

Assemble a kidding kit before you need it and store it in something easy to lug around. We pack ours in a Rubbermaid toolbox stool. It's roomy, has a lift-out tray for small items, is easy to carry to the barn, and is so much nicer to sit on than an overturned 5-gallon bucket. A good-size fishing tackle box is another fine option, as is a book bag or day pack to sling on your back. You will need:

- A bottle of 7 percent iodine (used to dip newborns' navels); since government regulations have made this difficult to find, you might have to substitute a weaker iodine solution, such as Betadine or Chlorhexidine (Nolvasan)
- A shot glass (to hold the navel-dipping fluid)
- Sharp scissors (for trimming an umbilical cord prior to dipping it); keep these scissors disinfected and stored in a sturdy Ziploc bag
- Dental floss (to tie off a bleeding umbilical cord in the event it's needed)
- A digital rectal thermometer

- A bulb syringe (the kind used to suck mucus out of a human infant's nostrils)
- A leg snare (a ready-made rubber one or a homemade version fashioned out of nylon cord)
- OB gloves
- Two large squeeze bottles of lube; we use SuperLube from Premier1 Supplies (see Resources)
- Betadine Scrub (for cleaning does prior to assisting)
- A sharp pocketknife (you never know when you need one)
- A hemostat (ditto)
- A lamb- and kid-carrying sling
- A halter or collar and lead
- Two flashlights (we like to have a backup in case the first one fails)
- Two clean terry-cloth towels

Prepare to Help If Needed

Most does kid without assistance, but you should be ready to help if the need arises.

Keep your fingernails clipped short and filed throughout kidding season; you won't have time for a manicure when trouble strikes, and you don't want to scratch and scrape sensitive tissues in the birthing canal.

Learn to determine the configuration of a poorly positioned kid and how to correct its delivery. If you think you might forget, photocopy and laminate a cheat sheet to stow in your kidding box.

Teach your fingers to "see." Collect a pile of plush toy animals. Chuck one in a paper bag. Without looking, stick your hand in the bag and figure out which parts you're feeling. Switch animals. Add another animal, then a third one; see if you can sort out triplets.

Program appropriate numbers into your cell phone. Be sure you can phone the vet or a goat-savvy friend or two at the touch of a button.

Install a baby monitor in your barn. Most does thrash around and many scream when they go into labor; an inexpensive audio monitor can save many moonlight strolls to the barn (video models are even nicer).

It's Almost Time

There are some telltale signs that your doe is about to give birth. You should learn and memorize these, so you can keep track of her progress.

Expanded udder. The average first-freshener's udder begins expanding about a month before kidding; a veteran doe's udder fills out anywhere from a few weeks to immediately prior to delivering her kids. Most does develop strutted udders a day or so before kidding. A strutted udder is so engorged that it's shiny and the teats jut out somewhat to the sides.

Soft pelvic ligaments. A stellar way to predict impending kidding is to monitor the pelvic ligaments in your doe's hindquarters. These ligaments attach at either side of the spine about midway between her hips and pin bones, then they angle toward the rear and away from the spine (one authority suggests you visualize a peace sign). To find them, slide your hand along the doe's spine, including the area an inch to either side. They're about as big around as a pencil and very firm, at least until secretion of the hormone relaxin softens them in preparation for giving birth. These ligaments become increasingly softer as kidding day approaches. When you can't feel them any longer, expect kids within the next 24 hours.

Steep rump. Relaxin causes additional structures in the pelvis to soften as well. The rump becomes increasingly steeper, both from hips to tail and as viewed from side to side. When this happens, the area along the spine

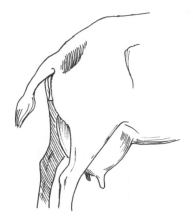

The ligaments in a doe's hindquarters become increasingly relaxed as kidding approaches.

A doe may express pain by gritting her teeth and throwing her head back as early labor contractions wrack her body.

seems to sink and the tail head rises. About 12 hours before kidding, you can grasp the spine at the tail head and almost touch your thumb to your fingers on the other side. It's a dramatic change and one you won't believe until you've felt it.

Swelled perineum. The perineum (the hairless area around a doe's vulva) sometimes bulges during the last month of pregnancy. About 24 hours before kidding, the bulge diminishes and the vulva becomes longer, flatter, and increasingly flaccid.

Mucus discharges. As the cervix begins to dilate, the cervical seal (wax plug) liquefies. When this occurs, does often discharge from their vulvas strings of mucus ranging from clear, thin goo to a thicker, opaque white substance to a thick, amber-colored discharge tinged with amniotic fluid.

Strange but True

Single-born bucklings are usually born a day or two sooner than single-born doelings.

• • •

Multiples are usually delivered two or three days sooner than singles.

• • •

When twins of opposite sexes are born, the male is nearly always delivered first.

• • •

Many does prefer male kids. If one kid from a litter is rejected, it's usually a doeling.

First-Stage Labor: Was That a Labor Pain?

During first-stage labor, which generally lasts 12 to 36 hours, mild to moderate uterine contractions may cause your doe to pause for a moment to stretch and raise her tail or to lie down briefly and hold her breath. These additional behaviors may also indicate first-stage labor:

- She may drift away from the herd to seek a private birthing spot, sometimes in the company of her dam or a daughter, sister, or best friend.
- She may become unusually affectionate or standoffish.
- She'll likely engage in nesting behavior by digging a depression in her stall or pen, lying down, getting up, circling, digging, and repeating the cycle over and over again.
- Her hind legs will move more loosely when she walks (as hormonal changes relax her pelvis, her rear leg angulation changes and this affects her gait) and she will move more slowly than usual.
- She may gaze into space with a wide-eyed, unblinking, faraway look on her face or yawn and stretch (stretching helps put her kids into birthing position).
- Some does search for their unborn kids while "talking" in a soft, low-pitched murmur (we call this her "mama voice").
- She might pant, stop eating, or grind her teeth (all indications that she's in pain).

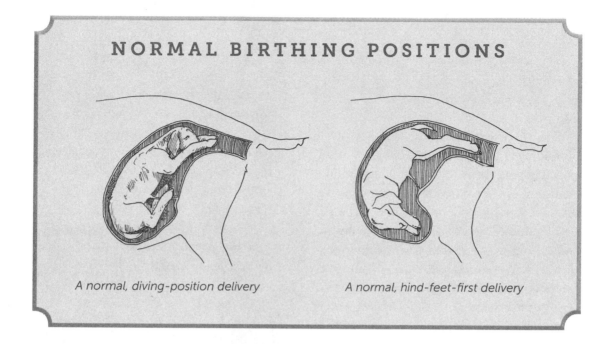

NORMAL BIRTHING POSITIONS

A normal, diving-position delivery

A normal, hind-feet-first delivery

Second-Stage Labor: Getting Down to Business

Eventually, second-stage labor begins. Your doe lies down and rolls onto her side when each contraction hits. She rides out the contraction and then rises and repeatedly repositions herself until she finds a spot she likes. Once she does, she may roll up onto her sternum between contractions, but she'll usually remain lying down. Some does, however, deliver standing up or in a squatting position, and that's okay, too.

The first thing you'll see at her vulva is a fluid-filled, water balloon–like sac called the chorion. The chorion is one of two separate sacs that enclose a developing fetus within its mother's womb (the other is the amnion). Either or both sacs might burst within your

doe or at the same time her kid is being delivered.

About 60 percent of the time, kids are delivered in a front-feet-first, diving position. Normally a hoof appears inside the chorion (or directly in the vulva if the chorion has already burst), followed by another hoof, and then the kid's nose tucked close to his knees. Once the head and shoulders are delivered, the rest of the kid usually slips out.

In a normal hind-feet-first delivery, two feet appear followed by two hocks. Because the umbilical cord is pressed against the rim of the doe's pelvis during this delivery, it's best to gently pull the kid once his hips appear. Otherwise, normal front-feet-first and hind-feet-first deliveries rarely require assistance.

Dealing with the Unexpected

Most does will have a normal delivery, in which case you won't need to intervene. In the rare event that there are complications, however, you should know what to do.

Your doe's first kid should be born within an hour after hard labor commences. Many producers swear by the 30-30-30 rule: allow 30 minutes after hard labor begins for the birth membranes to appear, another 30 for the first kid to be born, and 30 more minutes for second and subsequent kids to be delivered. If things aren't resolved in that time frame, your doe needs help.

If it appears she needs help, begin by examining your doe internally. Make sure your fingernails are short enough and smooth, and you've removed your watch and rings. Swab her vulva using warm water and mild soap or a product such as Betadine Scrub. Slip on an OB glove or scrub your hand and forearm with whatever you used to clean the doe. Then slather the glove or your hand and arm with lube. Now pinch your fingers together and gently ease your hand into her vulva. She will probably scream (and may keep screaming the entire time you're inside of her), so don't be surprised when she does.

Figure out which parts of the kid are present in the birth canal. Closing your eyes and moving your awareness to your fingertips help. If the kid's toe points upward and the big joint above it bends away from the direction the toe is pointing, it's a foreleg. If the toe points down and the major joint bends in the same direction, it's a hind leg.

Follow each leg to the shoulder or groin to make sure the parts you're feeling belong to the

What If the Doe Isn't Straining?

You have two options if your doe's contractions stop before a kid is born: call the vet or go in yourself. The best option is to call your vet without delay. He'll probably give her a shot of a drug called oxytocin, which will get her started again.

If the vet can't make it in time, slather on as much lube as you possibly can and, providing the kid is in a normal birthing position, slowly pull the kid. This is not easy and it's very dangerous for your doe. If you or the kid tears her, she'll die. Attempt this only as a last resort.

same kid. If they do and if you can manipulate the kid into a normal birthing position, do so. Cup your hand over sharp extremities like hooves and work carefully and deliberately. Then help your doe by pulling out the kid.

Goats aren't built like cows, so you can't pull a kid with tackle the way you'd pull a calf (the kid snare in your kidding kit, even though some people call it a kid puller, isn't designed to pull kids; it's used to fasten around legs so that you don't "lose" them while correcting dystocia). If you pull, use lots of lube. Simply grasp his legs, preferably above the pasterns but below the knees, and pull *only* while your doe is having a contraction. Don't pull straight back — pull down in a gentle curve toward your doe's hocks. In an emergency, you can also pull on his head.

If you have to pull one kid and there are others waiting to be born, pull them all. Your doe will be exhausted, so help her deliver them so she can rest. If you can't reposition the kid or kids in 10 minutes, stop trying and call your vet. Any time you go into a doe, you must follow the birth with a course of antibiotics. Ask your vet for advice.

Abnormal Birthing Positions

True breech presentation (butt first, hind legs tucked forward). Some does can give birth to kids in this position (our girl Bon Bon has done it not once but twice). However, many can't. If that's the case, call your vet. If you must reposition this kid by yourself, try to elevate your doe's hindquarters before you begin. Push the kid forward, work your hand past his body, and grasp one hock. Raise the hock up and rotate it out away from his body. While holding the leg in that position, use the little and ring fingers of the same hand to work the foot back and into normal position; repeat on the opposite side. The umbilical cord will be pinched, so pull this youngster as quickly as you safely can.

Head back. A small kid with its head bent back to its side can sometimes be pulled, but do your best to correct his position before you do. To correct this dystocia, attach a kid snare to the front legs so you don't lose them, then push the kid back as far as you can and bring his head into position.

Sometimes the kid's front legs are positioned properly but his head is bent forward and down. This is more difficult to correct. Try to correct it as you would for a head bent back, but call your vet and have him headed your way before you begin.

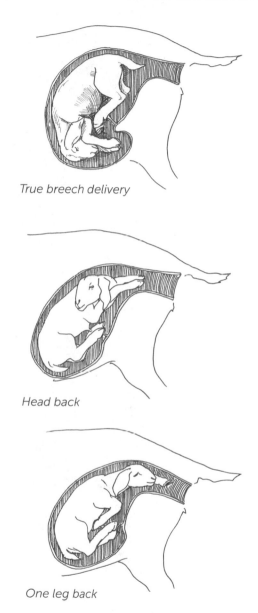

True breech delivery

Head back

One leg back

One leg back. Most does can deliver a kid in this position, but if yours can't, push her kid back just far enough to allow you to cup your hand around the offending hoof and gently pull it forward.

ABNORMAL BIRTHING POSITIONS
(CONTINUED)

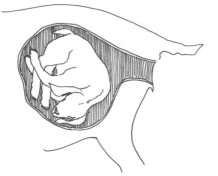

Crosswise

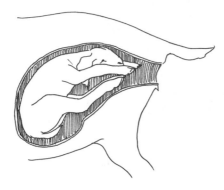

All four feet at once

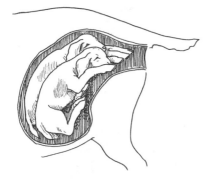

Twins coming out together

Crosswise. This is a hard one; call your vet. If no vet is available, push the kid back as far as you can (elevating your doe's hindquarters will help) and determine which end is closer to the birth canal, then begin manipulating that end into position. It is usually easiest to deliver these kids hind feet first.

All four legs at once. Attach a kid snare to one set of legs, making sure you have two of the same kind (front or hind) and then push the kid back as far as you can. Reposition the kid for either a diving position or hind-feet-first delivery, depending on which set of legs you have in the snare.

Twins coming out together. Attach a kid snare to two front legs of the same kid (follow the legs back to their source to make sure they're attached to the same kid). Push the other kid back as far as you can and bring the captured kid into the normal birthing position.

Twins coming out together with one reversed. Follow the same protocol as for twins coming out together, but since it's generally easier to do so, pull the reversed kid first. If both are reversed, first pull the kid closer to the birth canal.

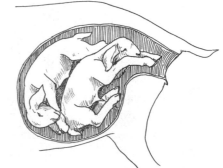

Twins coming out together with one reversed

Immediate Postdelivery Care: Here Be Kids!

When the first kid arrives, strip birthing fluids from his nose by firmly pressing and moving your fingers along the sides of his face. If he's struggling to breathe, use the bulb syringe from your kidding kit to remove excess fluid.

An alarming number of assisted kids appear to be dead when first pulled out, but their hearts are pumping; they simply haven't started to breathe. If yours is really struggling to breathe or not breathing at all, securely grip his hind legs between the hocks and pasterns with one hand, place your other hand behind his neck to support it, and swing him in a wide arc to jump-start his breathing. This usually sets a kid right. Tickling the inside of a kid's nostrils with a piece of straw or hay works, too. Don't give up. If you keep stimulating the kid, chances are he'll start to breathe and be perfectly all right.

Once he's breathing, place the kid in front of his dam so she can clean him. It's the taste and scent of her newborn kid that creates her maternal bond. At some point she may leave him to deliver another kid. This is normal. Simply place both kids in front of her after the second has arrived.

Once all kids are accounted for (if in doubt, go in and check), think "snip, dip, strip, and sip."

1. Snip the umbilical cord to a manageable length if it's overly long, about 1½ inches (4 cm) is just right.

2. Dip the cord in iodine. Fill a shot glass, film canister, or dairy cow teat dip cup with iodine, hold the container to the kid's belly so the cord is completely submerged, then tip the kid back for full coverage.

3. Strip your doe's teats to make certain they aren't plugged and that she indeed has milk.

4. Make sure the kids sip their first meal of colostrum within an hour or so after they're born.

In a best-case situation, your doe's afterbirth will pass within a few hours after her final kid arrives. Most does eat the afterbirth if you let them, but this a choking hazard, so it's best to remove membranes as soon as they're delivered. If the afterbirth isn't delivered within 12 hours (you'll know because bits will still be dangling from her vulva), call your vet.

Swinging a seemingly stillborn kid may be all that's needed to jump-start his lungs and get him breathing.

Never pull on birthing tissues for any reason; doing so can cause the doe to prolapse.

Neonatal kids are wired to seek food in dark places, such as armpits and groins. Help weak or disoriented kids by holding them near their dam's udder, but let them find the teat by themselves (most kids resist a teat placed directly into their mouth).

Kids butt their dam's udder to facilitate milk letdown. A rapidly wagging tail means a kid is suckling milk. After feeding, contented kids snooze. A kid who cries, noisily suckles a lot, or constantly fiddles with his dam's udder isn't getting enough to eat. You may have to bottle-feed this baby.

Nannyberries

THE GOAT IN OLD IRELAND

The Goat, *Gabhar*, the *Capra hircus*, has in all probability existed in Ireland from the earliest period of its inhabitation, and the head, horns, and other bones of this animal have been found not only in crannogs but also in artificial caves and in the stone passages and vaulted chambers in ancient raths.

As the goat always gives way to the sheep in the progress of civilization, except in those regions which, from their peculiarly mountainous and rocky nature, must remain its natural habitat, we find this animal gradually disappearing in many localities formerly celebrated for it in Ireland. It is seldom alluded to in Irish writings of antiquity, and is not enumerated among the animals which were given as tributes to the kings of Erinn. . . .

The old Irish goat was small, in some instances white, but more usually of an iron-gray color. Many localities throughout the country, hills, rocks, and mountains, derive their names from goats, such as *Keam-a-gower*, the goat's path, in the west of the county of Cork; *Lisnanagabhar*, the goat's fort, in the county of Monaghan; and the celebrated pass in Achill Island, called *Minaun*, or kid's path.

St. Patrick had two buck-goats, which he employed for carrying water. An account of them will be found in Colgan's *Trias Thaumaturga*. They were stolen by three thieves of the Ui-Torra, in the territory of Hy-Meith-tire, in the now county of Monaghan; but the saint received information that enabled him to detect the thieves, who declared that they had not stolen the goats. Patrick, however, it is stated, worked a miracle on the occasion, and caused the animals, which they had killed and eaten, to bleat from their bellies, and he prayed that the descendants of the thieves should, throughout all time, be distinguished by producing and wearing on their chins beards similar to those of buck-goats.

— from *Proceedings of the Royal Irish Academy, vol. 7 (1862)*

Don't use does of any breed to foster lambs. Some does will happily adopt lambs, but lambs can damage their sensitive udders. While kids do bunt their dams' udders to facilitate milk letdown, lambs are wired to bunt savagely hard. It's better to milk your goat and use her milk to bottle-feed your lambs.

Handle your kids often so they grow up trusting humans, and check them over at least twice a day. Boisterous kids injure themselves, and it doesn't take long for a kid to die from life-threatening kidhood diseases (revisit chapter 10 for particulars).

Treating Mama Right

Milking does require proper feed. Talk to your county Extension agent or an experienced goat milker in your locale to formulate a diet based on local feeds, or feed alfalfa hay and commercial grain supplement such as Purina Goat Chow, following the instructions on the bag.

Preventing and Treating Mastitis

Mastitis is defined as inflammation of the udder. It's caused by a wide array of bacterial and staph agents. Both intramammary infusions (introduced directly into the teat canal) and systemic (body-wide) antibiotics are generally used to treat this serious disease. If you think your doe might have mastitis, have your vet culture a milk sample to find out which infective agents are involved; that way your vet will know which medications to prescribe.

Subclinical mastitis symptoms are subtle; they can be detected only in the strip cup or through testing. Acute mastitis symptoms include a hot, swollen, hard or lumpy udder; extreme pain; lameness; loss of appetite; fever;

decreased milk production; clumps, strings, or blood in the milk; and watery, strange-looking milk. Gangrene mastitis presents as a bruised-looking, extremely swollen and painful udder that turns blue as infection takes hold.

Mastitis can be triggered by substandard milking hygiene and delayed milking (for some does, just a few hours' delay can cause a bout); udder injuries; stress; and milk buildup after does wean their kids. Untreated, it can progress to a serious form called gangrene mastitis, and, ultimately, death.

Good news: mastitis is not an everyday thing. If you take good care of your doe and milk her under sanitary conditions, you'll probably never see mastitis in your barn. And if it does strike, you can clear up problems before they become acute by conscientiously testing your goat's milk at weekly intervals and treating subclinical mastitis.

To test your milk for mastitis, you need a California Mastitis Test (CMT) kit, which contains a plastic paddle with four compartments (the kit was designed for testing cows); a bottle of CMT solution, which you'll reconstitute according to package instructions; and a test results card, which shows precisely what you're looking for as you test.

You'll perform this test just before milking. Squirt 1 teaspoon (2 cc) of milk into each of two compartments of the paddle. Add 3 cc of reconstituted CMT solution to each compartment, then rotate the paddle in a circular motion to mix the compartments' contents. Ten seconds after adding the solution, read the results (visible reaction disintegrates after about 20 seconds, so you have to act fast).

If there is little or no thickening of the mixture, the test is negative (the CMT test card was developed for testing cow's milk, so despite what the instructions say, healthy goat's milk may show the very slightest hint of thickening). Definite thickening indicates the possibility of mastitis; if the mixture gels, mastitis is present.

If the CMT indicates mastitis or if you just aren't certain, take a sample of milk to your vet to be cultured. Don't wait to see if things get worse. This isn't a disease you can ignore.

Raising Bottle Kids

If you want a goat that will live to please you, raise a bottle kid. They're adorable, affectionate, and sure to steal your heart. Even as adults, they look to you as their leader, making them the best companions, packers, and harness goats bar none.

It isn't difficult to raise a bottle kid, but it's also not a lark. When you assume responsibility for a newborn of any species, you must be prepared mentally, materially, and emotionally for what lies ahead. For 10 to 12 weeks

When bottle-feeding a kid, hold the bottle at an angle, just enough so that the nipple is filled with milk.

(or more if you choose), a bottle kid will be entirely dependent on you.

You'll need to devote time and energy to your kid. For the first few weeks you'll be feeding four to six times every 24 hours, and some of those feedings will fall in the dead of the night. Forget about ignoring the clock; your kid needs meals delivered on time. He'll also crave your attention. If you can't spend a lot of time with him, consider adopting two bottle kids so they'll have each other for company when you're not around. Two will love you just as much as one, and they're double the fun.

You'll also need to put some money into your kid's dietary and physical needs. You mustn't skimp on milk or formula. There are several accepted ways to feed bottle kids (we'll talk about that in a moment) but none is dirt cheap. Baby goats have specific nutritional needs; if you don't meet them, they often fail to thrive. Your kid will also need a safe, warm, and tidy place to live. Kids can be fragile. They don't "do" drafts, faulty ventilation, extreme cold or heat, crowding, or muck and dirt.

Although not all bottle babies get sick or injured, depending on where you get your kid and how you manage him, he might get hurt or contract any of several serious kidhood maladies. If that happens, he'll require veterinary care. Unless you're willing to provide it (as well as any home nursing a sick or injured kid might require), bypass a bottle baby and start with a sturdier, older kid or a grown-up goat.

Supply Time: Be Prepared!

Before bringing home your bottle kid, assemble the supplies and equipment you'll need to raise him. Time is of the essence with a newborn, and you can't go shopping with a goat tucked under your arm.

Colostrum

Newborn kids need colostrum — the thick, yellow, nutrient- and antibody-rich milk does produce for the first few days of each lactation. Because kids are born without disease resistance of their own, and since they absorb antibodies through the colostrum for only 12 to 36 hours after birth, it's vitally important that every kid ingest 10 to 15 percent of his body weight in colostrum during that day and a half, preferably 50 percent of his colostrum needs within 4 to 8 hours of birth.

If you buy or are given a newborn, ask the breeder if your goat received his essential nutrients. Responsible goat sellers make certain kids get those vital first meals, whether by nursing their dams, being given bottle- or tube-fed thawed colostrum from the freezer, or taking a quality colostrum substitute. But it never hurts to double-check. If your kid didn't ingest that elixir of life, he'll be at risk for disease until his own immune system kicks in, at 7 to 8 weeks of age.

If the breeder provides you with fresh goat, sheep, or cow colostrum, ask if the donors are CAE- and Johne's-free. If not, heat-treat the fluid before feeding it to your kid. Heat-treating colostrum is the essence of simplicity, but it's exacting work: gently warm the fluid to 133°F (56°C) in a double boiler (use a thermometer — it's important) and hold it at that temperature for 1 hour, stirring constantly.

Err on the side of caution. Before purchasing a kid, buy at least 16 ounces of colostrum from a goat or cattle dairy farm. Heat-treat, bag, and freeze it as soon as you can. It's most convenient to divide the colostrum into individual 2-ounce feedings and store them in resealable sandwich bags. Squeeze out excess air, seal, and store the baggies in a larger resealable freezer bag (for example, Ziploc bags). Toss the larger bag in a freezer that is *not* frost-free (the constant thaw and freeze cycles in a frost-free freezer compromise colostrum integrity). Properly processed and stored, the antibodies in frozen colostrum remain viable for at least a year. Thaw each feeding at room temperature or immerse the baggie in warm water. Don't heat it over the stove or in the microwave; excess heat and microwaves destroy vital antibodies.

If you can't find colostrum, the next best thing is an injection of Goat Serum Concentrate, an injectable immunotherapy product manufactured by Sera (see Resources). Injected subcutaneously (not into muscle) as soon after birth as possible, it protects newborns from a broad spectrum of viral and bacterial agents. At roughly $4 per dose, it's cheap insurance.

Farm stores and livestock supply catalogs also carry an array of colostrum boosters and replacers based on freeze-dried, whey-based, or serum-derived cow colostrum. Whether they help is debatable. If you use one, you may want to choose bovine serum–based supplements such as Colostrix, Immustart, and Lifeline — Colorado State University research indicates these work best.

• • • • • • • • • • • • • •

One goat cannot carry another goat's tail.

~ Nigerian proverb

• • • • • • • • • • • • • •

This child, shown in a turn-of-the-nineteenth-century vintage photograph, knows how to feed a goat! She's holding the bottle at the perfect angle for her kid.

Milk or Formula

The best food for kids is fresh goat's milk — but only if it comes from CAE-free herds or is pasteurized prior to feeding.

Pasteurized fluid or powdered goat's milk from the supermarket or health food store works, but it's expensive. If you're purchasing milk from a store, a more economical option is pasteurized, homogenized, whole-fat cow's milk. Some folks feed it straight from the jug, but it's better to boost its fat content by adding dairy half-and-half (1 part half-and-half to 5 parts whole milk). Young goats thrive on this combination, and it's a snap to store and prepare.

Milk replacers formulated for goat kids work, too; however, choose a good one. Inexpensive milk replacers are frequently based on soy rather than milk protein; kids can't digest it and they'll fail to thrive. Better products such as Purina, Land O' Lakes, Advance Kid milk replacers, and Merrick's Super Kid are excellent choices. Despite what their labels claim, *never* use one-kind-fits-all milk replacers or products formulated for the young of other species. Always opt for milk protein–based products labeled specifically for goats.

It's best to stick to the same milk or formula for as long as you bottle-feed your kid. Abruptly switching from milk to milk replacer or from one brand of milk replacer to another upsets a kid's delicate tummy. The result: diarrhea, also known as scours. It's not pretty, and it's dangerous. Diarrhea leads to dehydration and serious problems such as enterotoxemia, so if you must switch, do it gradually over the course of several days.

Bottles and Nipples

The best kid-feeding device we've used is the genuine Pritchard flutter-valve nipple. This funny-looking soft, red rubber nipple works well for weak or newborn kids, yet is tough enough to withstand the chewing efforts of older, lusty feeders. The Pritchard nipple's yellow plastic base can be screwed onto any 28 mm glass or plastic bottle (20-ounce soft-plastic soda pop bottles are ideal). A metal ball bearing inset in a hole in its base neatly regulates air and milk flow so kids neither suck excess air nor choke.

To prepare a Pritchard nipple for a newborn, carefully snip off the tiniest bit of the tip, leaving a hole just large enough for him to nurse; for other types of nipples, slightly enlarge the hole using a hot needle. Don't make the hole too large; kids who nurse too quickly get fluid in their lungs. You want them to suck, not guzzle.

Among other kid-friendly nipples are the soft isoprene nipple supplied with the Non-Vac Nurser Set and the Controlled Flow Pop Bottle Nipple, both sold through Hoegger Goat Supply (see Resources).

Or go with what you may already know: bottles and nipples designed for human infants work for goat kids too. And when your kid chews up an everyday infant nurser nipple, you can drive to the store and pick up more.

What you don't want is the somewhat hard, black rubber lamb nipple available through livestock catalogs and at local farm stores. Small kids can't suck vigorously enough to nurse from them, while older kids jerk them off feeding bottles and choke.

Buy enough nipples to last the duration of bottle feeding. Kids are creatures of habit and it's hard to convince them to switch to a different type of nipple, even in a pinch. Save yourself aggravation and grief, and stockpile nipples, especially if you can't buy replacements close to home.

Buy a bottlebrush to scrub the bottles, a calibrated measuring cup for mixing formula, a plastic funnel, and a mild disinfectant solution to keep everything squeaky clean.

And if you're game to take on a sick or weak baby, buy tube-feeding apparatus as well. You'll need a 60 cc disposable syringe and a soft plastic or rubber stomach-feeding tube to feed him until he's strong enough to nurse from a bottle (see page 189 for how to do it).

Bringing a Kid Home

Consider keeping your kid in your home for 4 to 6 weeks. When housed in a roomy dog kennel with a flip-open top, a folding canine exercise pen, or an old-style wooden playpen, and kept clean and dry, a kid has surprisingly little odor. Diapered, he can frisk through the house and entertain you. It certainly beats watching TV.

A kid kept indoors bonds with his human family, learns to interact with other household pets, and is just down the hall when his 3 A.M. bottle comes due. That counts for a lot when the mercury dips below freezing and your barn is a long trudge from the house.

One warning when raising house kids: Kids are adept climbers and inveterate chewers, so don't leave them unsupervised for any length of time.

Bed and Clothe Him Right

Bed your kid's indoor crib with old blankets (half of a fluffy, double- or queen-size blanket folded in thirds pads a roomy dog crate). During the first week or two he'll make yellowish pudding-consistency poo, so you'll need to change his blankets fairly often. As he begins nibbling grass or hay, he'll begin ejecting nice, firm "nannyberries" you can scoop up with a tissue or paper towel; then you'll change bedding only as it becomes damp. He'll also need at least two tip-proof dog bowls or crate cups: one for water and one (or more) for dry feed.

Much of the fun of raising an indoor kid is letting him run rampant through the house. However, if puddles and pellets are a turnoff, diaper your kid during indoor outings. Doelings are easy: cotton or disposable diapers for human infants of comparable size neatly do the trick (be sure to snip out a hole for baby's tail).

In addition to a diaper, males can be fitted with the sort of washable, around-the-middle band designed for incontinent older dogs (most pet stores carry them). Doelings sometimes train themselves not to urinate while loose in the house. If your doeling is one of these, let her romp undiapered with easy access to her crate and simply suck up her nannyberries with a Dustbuster.

Choose the Right Companion

If you're raising a pair of kids, they'll reassure each other, but a single kid needs a large, soft toy to cuddle up to. Choose a pal that can be laundered and quickly dried — plush toys designed for dogs work well. Sturdy human infant chew toys or a bell will delight him, too,

especially if suspended by a sturdy cord at his eye level.

As for dogs, most are good with kids as long as you're watching. However, canines, with the exception of trained livestock guardian dogs, are wired to pursue things that flee. Goat play can be misinterpreted by the gentlest dog, so don't leave Fido and a kid unsupervised, in or outside of your home.

Transition to the Outdoors

As your kid matures, you'll probably want to move him to your barn, garage, or other outdoor housing. Make certain his outdoor quarters are warm, safe, well ventilated, and draft-free. If it's cold outside, don't warm him with a heat lamp; a lamp can spark a tragic fire. Instead, clothe him to keep him warm.

A cardigan-style human baby sweater makes a fine kid cover; place your kid's legs through the sleeves and button the sweater up his back. Or choose a dog garment or a jacket specifically tailored for lambs and kids. And make sure he has a companion in a nearby area — be it another goat, a dog or cat, some chickens, a sheep, or a horse. He'll be a lonely baby if you don't.

Additional Health Care How-to

Tube-Feeding Baby

If your kid can't suck, you'll have to stomach-tube-feed him, which sounds a lot scarier than it is. You'll need an empty 60 cc syringe (60 cc equals 2 fluid ounces, the correct dose per feeding for most newborn kids) and a soft plastic feeding tube. If you didn't buy this kit in advance, your veterinarian may have the components to sell you. Just the syringe and 28 inches (71 cm) of soft, unused, ¼-inch plastic tubing from the hardware store or a sterile catheter from a hospital or nursing home will suffice in a pinch.

While the following process seems time-consuming, dangerous, and difficult, it's not. After a few trial runs, it will take you just a few minutes to tube-feed from start to finish, and it's safer than dribbling colostrum into a weak baby's mouth.

If you can recruit a helper for this procedure, so much the better.

1. Place a premeasured amount of colostrum in a glass measuring cup and remove the barrel from the syringe.

2. Lay the tube along the kid's side and measure from just in front of his last rib to his mouth. Mark that point with tape or indelible marker.

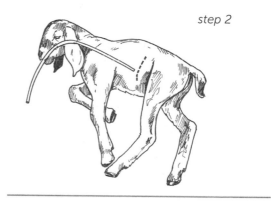

step 2

3. Place the kid in your lap on either his left or right side (he can sit up if he's strong enough), facing away from you. Gently insert your thumb between his lower teeth and upper dental palate to open his mouth.

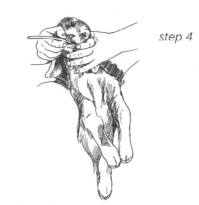

step 4

4. Keeping a finger in his mouth, carefully insert the tube along the side of his mouth and feed it slowly, allowing the kid to swallow it if he can. Keep passing the tube until the mark you made on it is even with his mouth.

5. It's important that you don't insert the tube into his trachea (windpipe), as feeding fluid into his lungs will trigger pneumonia and kill him. For the same reason, syringing colostrum into his mouth is a bad idea. A kid's esophagus is soft; if, keeping your fingers on his throat, you can feel the tube descending, you're likely okay. If it does enter his trachea, the kid will struggle violently and the tube will halt midway to your mark. Gently pull it out and start again.

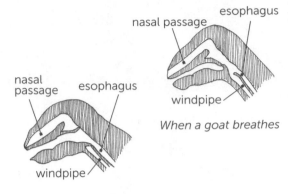

When a goat breathes

When a goat swallows

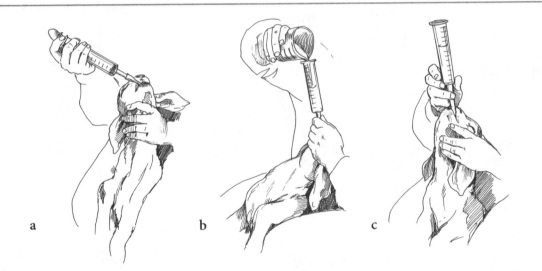

a

b

c

6. With the tube in place and while supporting the kid as best you can (this is where a helper comes in handy), attach the barrel of the syringe (**a**), add the colostrum (**b**), then hold the syringe up high and allow the fluid to be fed by gravity into the kid's stomach (**c**).

7. Allow all of the precious fluid to reach its mark, then firmly pinch the tube and draw it back out. Don't hesitate midway and don't forget to pinch; you don't want to spill liquid into his lungs.

Giving an Intramuscular Injection

Some antibiotics are given intramuscularly (IM; into a major muscle mass). To give an intramuscular injection, use a 1- to 1½-inch, 18- or 20-gauge needle. Some antibiotics are very thick and the carriers used in their manufacture make these injections sting a great deal; for these, choose a fat, 16-gauge needle to inject the fluid quickly before your goat objects.

Intramuscular injections are usually given into the thick muscles of a goat's neck.

1. Quickly but smoothly jab the needle deep into the muscle mass.

2. Aspirate (pull back on) the plunger ½ inch (1¼ cm) to be sure you have not hit a vein. If blood rushes into the syringe, pull out the needle, taking care not to inject any drug or vaccine as you do, and try another injection site.

Drenching Your Goat

Some medicines will be in paste or liquid form, so it's a good idea to learn how to administer them to your goat (called drenching). Drenches can be administered using a catheter-tip syringe (not the kind you use to give shots) and a turkey baster, but the most efficient way is to use a dose syringe. Follow these instructions.

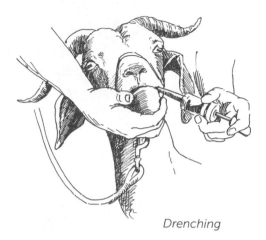

Drenching

1. Back your goat into a corner so he can't escape, restrain him (straddle his back, facing forward, if you're tall enough not to be taken for a ride), and, using one hand under his chin, slightly elevate his head — just enough to allow gravity to help you a little bit.

2. Insert the nozzle of the syringe between your goat's back teeth and his cheek (this way the goat is less likely to suck fluid into his lungs) and slowly depress the plunger, giving him ample time to swallow.

When giving a goat semisolid substances such as paste-type dewormers and gelled medications, deposit the substance as far back on the goat's tongue as you can reach.

3. Keep your goat's nose slightly elevated until he visibly swallows. And be careful: if you stick your fingers between his back teeth, you're likely to be bitten, and a goat's back teeth are razor sharp!

Pilling Your Goat

Ask your vet if the pill he prescribes must be given whole. If it doesn't need to be, thoroughly smash the pill using a mortar and pestle or two spoons (if it's small), or pop it into a paper bag and smash it with a hammer (if it's big). Wear gloves; never handle fragments or powdered pills with bare hands. Next, dissolve the dust in liquid (water, fruit juice, or milk) to give as a liquid drench or stir it into yogurt and give it like paste. The latter is the easier of the two.

If the pill must remain whole, give it with a balling gun. This is a goat-size plastic device designed to propel a pill down a goat's throat. Follow these instructions.

1. Coat the pill with molasses or yogurt, place it in the balling gun, then lay aside the assembly.

2. Secure the goat as though you were giving him a drench.

3. Pry his mouth open (watch your fingers!), place the loaded balling gun far enough in his mouth to deposit the pill on the back of his tongue, and raise his head to a 45-degree angle.

4. Depress the plunger, pull out the balling gun, stroke his throat, and hope. If you're lucky, he'll swallow the pill. If he doesn't, pick it up and try again.

A balling gun and some pills

Resources

Dreamgoat Annie
www.dreamgoatannie.com
The author's website; lots of additional resources

Clicker Training

Clicker Solutions
www.clickersolutions.com

Karen Pryor Clickertraining
www.clickertraining.com
Clickers, targets, training gear, and books

On-Target Training with Shawna Karrasch
http://on-target-training.com
Clickers, targets, books, and information

Must-See Videos on YouTube

Clicker Sheep
www.youtube.com/watch?v=YKnSls3zz9A
Inspiring!

Spotty's Tricks
www.youtube.com/watch?v=P7Y-gmSIR90
Even more inspiring!

Dairying and Cheese Making

Information

A Campaign for Real Milk
The Weston A. Price Foundation
www.realmilk.com

Start a Grade A or Grade B Goat Dairy
American Dairy Goat Association
http://adga.org/StartDairy.htm
How to contact your state's dairy regulatory division

Summary of Raw Milk Statutes and Administrative Codes
A Campaign for Real Milk
www.realmilk.com/milk-laws-1.html

Cheese Making Supplies

New England Cheesemaking Supply Company
Ashfield, Massachusetts
413-628-3808
www.cheesemaking.com

Steve Shapson's The Cheese Maker
Cedarburg Homebrew and Wine
Cedarburg, Wisconsin
414-745-5483
www.thecheesemaker.com

Driving

Harnesses, Carts, and Wagons

Buggy-n-On
Peachland, North Carolina
704-272-9014
www.buggynon.com
Biothane (synthetic leather) goat harnesses

Caprine Supply
De Soto, Kansas
913-585-1191
www.caprinesupply.com
Wagon and nylon goat harnesses

DW Farms
Adam Black, Goat Harnessmaker
Dresden, Ohio
740-683-4628
http://workinggoats.tripod.com
High-quality nylon harnesses

Gentle Spirit Behavior and Training for Llamas and Alpacas
Cathy Spaulding
www.gentlespiritllamas.com
Adjustable halters, perfect for goats

H & S Horse Cart Co., Ltd.
sale@hscart.com
www.hscart.com
Goat/dog/miniature horse carts with removable third wheel

Hoegger Goat Supply
Fayetteville, Georgia
800-221-4628
www.hoeggergoatsupply.com
Carts, wagons, and several types of goat harness

Nikki's Pony Express
Middlebury, Connecticut
203-758-2354
www.nikkisponyexpress.net
Carts and goat harnesses

Northern Tool + Equipment Co.
Burnsville, Minnesota
800-221-0516
www.northerntool.com
A good place to mail-order utility wagons to convert for goats

Quality Llama Products Inc. & Alternative Livestock Supply
Lebanon, Oregon
800-638-4689
www.llamaproducts.com
Nylon goat harnesses and goat driving halters

Ravenwood Farm
Hartsville, Indiana
812-546-6128
www.ravenwoodfarm.com
Custom-made leather and Beta
(synthetic leather) goat harnesses

Tractor Supply Co.
Brentwood, Tennessee
877-718-6750
www.tractorsupply.com
A good source for utility wagons to
convert for goats

Wilczek Woodworks
Littleton, New Hampshire
603-444-0824
www.wilczekwoodworks.com
Stunning wooden dog carts and wagons that are readily adapted for goats

Working Goats
www.workinggoats.com
Nylon wagon harnesses for goats

Information

4-H Harness Goat Project
Cornell University
*www.ansci.cornell.edu/4H/goats/
harness_goat.pdf*

**Converting a Child's Wagon
to a Cart**
Carting with Your Dog
*www.cartingwithyourdog.com/
cwagonconvert.html*

**Goat Tracks Magazine: Journal of
the Working Goat**
www.goattracksmagazine.com

Hounds of Caid
http://houndsofcaid.sca-caid.org
Plans to build period vehicles for large
dogs that are adaptable for use with
goats; click on *Activities* and then
Carting

Working Goats
www.workinggoats.com

Organizations

**American Harness Goat
Association**
Yelm, Washington
*www.goattracksmagazine.com/harness.
html*

**Canadian Pack and Harness
Goat Association**
Goats Across Canada magazine
McBride, British Columbia
250-569-7406
www.goatsacrosscanada.ca
Click on *Canadian Pack and Harness
Goat Association*

Harness Goat Society
Worcester, United Kingdom
harnessgoatsociety.uk@virgin.net
www.harnessgoats.co.uk

Goat-Care Information

Amber Waves Pygmy Goats
http://amberwavespygmygoats.com

The Biology of the Goat
www.imagecyte.com/goats.html

**E (Kika) de la Garza Institute for
Goat Research**
Langston University Goat and
Research Extension
www2.luresext.edu/goats
Don't miss the interactive nutrient-
requirement calculator and the free
Web-based meat goat producers'
course (you'll learn a lot, no matter
what type of goats you own). Click
on *Library* to access an array of great
resources, including step-by-step
instructions for conducting your own
fecal exams.

Fias Co Farm
http://fiascofarm.com

Goat World
www.goatworld.com

Kinne's Minis
http://kinne.net

Maryland Small Ruminant Page
University of Maryland Extension
www.sheepandgoat.com

Meat Goat Home Study Course
Penn State Cooperative Extension,
Bedford County
*http://bedford.extension.psu.edu/
agriculture/goat/Goat%20Lessons.htm*
It's *very* good no matter what type
of goats you keep, and online lesson
materials are free.

Shepherd's Notebook
http://mdsheepgoat.blogspot.com
Blog written by Susan Schoenian,
sheep and goat specialist for the
University of Maryland's Cooperative
Extension

Goat Fairs

International Goat Days Festival
Millington, Tennessee
www.internationalgoatdays.com

Puck Fair
Killorglin, Ireland
www.puckfair.ie

Goat Organizations

Breeds

UNITED STATES

Alpines International Club
Silt, Colorado
970-876-2738
www.alpinesinternationalclub.com

**American Angora Goat Breeders
Association**
Rocksprings, Texas
830-683-4483
www.aagba.org

American Boer Goat Association
San Angelo, Texas
325-486-2242
www.abga.org

American Dairy Goat Association
Spindale, North Carolina
828-286-3801
www.adga.org

American Goat Society
Pipe Creek, Texas
830-535-4247
www.americangoatsociety.com

American Kiko Goat Association
Killeen, Texas
254-423-5914
www.kikogoats.com

American LaMancha Club
Redding, California
alc@lamanchas.com
www.lamanchas.com

American Miniature Dairy Goat Association
Kennewick, Washington
509-591-4256
www.miniaturedairygoats.com

American Nigerian Dwarf Dairy Goat Association
Oolitic, Indiana
www.andda.org

American Nigora Goat Breeders Association
McMinnville, Oregon
nigoragoat_info@yahoo.com
http://nigoragoats.homestead.com

American Silky Fainting Goat Registry
Salina, Kansas
registar@asfgr.com
www.asfgr.com

Arapawa Goat Breeders — USA
Rehoboth, Massachusetts
508-252-9469
www.arapawagoat.org

Colored Angora Goat Breeders Association
Rogue River, Oregon
541-582-3705
info@cagba.org
www.cagba.org

Eastern Cashmere Association
Staunton, Virginia
scott.pasini3@wildblue.net
www.easterncashmereassociation.org

International Boer Goat Association, Inc.
Whitewright, Texas
903-364-5735
www.intlboergoat.org

International Dairy Goat Registry
Milo, Missouri
417-944-2455
www.goat-idgr.com

International Fainting Goat Association
Bennet, Nebraska
402-782-2089
www.faintinggoat.com

International Kiko Goat Association
Bluff City, Tennessee
888-538-4279
www.theikga.org

International Nubian Breeders Association
secretary@i-n-b-a.org
www.i-n-b-a.org

International Sable Breeders Association
www.sabledairygoats.com

Kinder Goat Breeders Association
Miami, Missouri
www.kindergoatbreeders.com

Miniature Dairy Goat Association
Kennewick, Washington
509-591-4256
www.miniaturedairygoats.com

The Miniature Goat Registry
Jamul, California
619-669-9978
www.tmgronline.org

Miniature Silky Fainting Goat Association LLC
Lignum, Virginia
540-423-9193
www.msfgaregistry.com

Myotonic Goat Registry
Adger, Alabama
205-425-5954
www.myotonicgoatregistry.net

National Miniature Goat Association
Bloomington, Indiana
www.nmga.net

National Mini-Nubian Breeders Club
Boyertown, Pennsylvania
www.mininubians.com

National Pygmy Goat Association
Snohomish, Washington
425-334-6506
www.npga-pygmy.com

National Saanen Breeders Association
Santa Cruz, New Mexico
Secretary-Treasurer@nationalsaanen breeders.com
http://nationalsaanenbreeders.com

National Toggenburg Club
Buhl, Idaho
208-543-8824
http://nationaltoggclub.org

Navajo Angora Goat Record
Del Norte, Colorado
505-400-9949
www.navajoangoragoat.org

Nigerian Dwarf Goat Association
Wilhoit, Arizona
www.ndga.org

North American Savanna Association
Hallsville, Missouri
573-696-2550
www.savannahassociation.com

Oberhasli Breeders of America
secretary@oberhasli.net
http://oberhasli.net

Pedigree International LLC
Humansville, Missouri
417-754-1155
www.pedigreeinternational.com
Registers Cashmere, Genemaster, Myotonic, Savanna, Spanish, Tennessee Meat Goat, and TexMaster

Pygora Breeders Association
Lysander, New York
315-678-2812
www.pygoragoats.org

San Clemente Island Goat
Association
  The Plains, Virginia
  540-687-8871
  www.scigoats.org

Spanish Goat Association
  The Plains, Virginia
  540-687-8871
  www.spanishgoats.org

United States Boer Goat
Association
  Spicewood, Texas
  866-668-7242
  www.usbga.org

CANADA

Canadian Goat Society
  Ottawa, Ontario
  613-731-9894
  www.goats.ca
  Registers Alpines, Angoras, LaManchas, Nigerian Dwarfs, Nubians,
  Oberhaslis, Pygmy Goats, Saanens,
  and Toggenburgs

Canadian Meat Goat Association
  Annaheim, Saskatchewan
  306-598-4322
  http://canadianmeatgoat.com
  Registers Boers and Boer crosses

UNITED KINGDOM

Anglo-Nubian Breed Society
  North Yorkshire
  www.anglo-nubian.org.uk

Bagot Goat Society
  Powys, Wales
  info@bagotgoats.co.uk
  www.bagotgoats.co.uk

Bilberry Goat Heritage Trust
  Waterford
  bilberrygoatheritagetrust@hotmail.com
  www.bilberrygoatheritagetrust.com

British Alpine Breed Society
  Wiltshire
  +44-01249-716350
  www.britishalpines.co.uk

British Angora Goat Society
  Warwickshire
  +44-01789-841930
  www.britishangoragoats.org.uk

British Boer Goat Society
  Devon
  info@britishboergoatsociety.co.uk
  www.britishboergoatsociety.co.uk

British Feral Goat Research Group
  feralgoatuk@yahoo.co.uk
  http://britishferalgoat.freeservers.com

British Goat Society
  +44-01434-240866
  Northumberland
  www.allgoats.com

British Toggenburg Society
  Kent
  +44-01795-866202
  www.britishtoggenburgs.co.uk

English Goat Breeders Association
  +44-01296-640842
  www.egba.org.uk

Golden Guernsey Goat Society
  enquiries@goldenguernseygoat.org.uk
  www.goldenguernseygoat.org.uk

Lynton Feral Goat Preservation
Society
  www.lyntongoats.org.uk

Old English Goat Society
  Ingelton
  +44-015242-41392
  www.oldenglishgoats.org.uk

Pygmy Goat Club
  Surrey
  +44-01372-818039
  www.pygmygoatclub.org

Toggenburg Breeders Society
  Worcestershire
  +44-01386-761892
  www.toggenburg-breedersociety.co.uk

AUSTRALIA AND NEW ZEALAND

Australian Association for Dairy
Goats
  Cambrai
  *registrar@theaustralianassociationfordairy
  goats.com*
  *www.theaustralianassociationfordairy
  goats.com*

Australian Cashmere Growers
Association Ltd.
  St. Lucia South
  +61-04-8875-6176
  http://acga.org.au

Australian Miniature Goat
Association Inc.
  Murwillumbah
  +61-04-1965-5348
  www.australianminiaturegoat.com.au

Boer Goat Breeders of Association
of Australia
  ABRI, University of New England
  Armidale
  +61-02-6773-5177
  www.australianboergoat.com.au

Dairy Goat Society of Australia
  Traralgon
  +61-03-5176-0388
  www.dairygoats.org.au

Miniature Goat Breeders
Association of Australia
  Boyland
  +61-07-5543-4625
  *www.miniaturegoatbreedersassociation.
  com.au*

New Zealand Boer Goat Breeders
Association
  boergoats@nzboer.co.nz
  www.nzbgba.co.nz

New Zealand Dairy Goat
Association
  info@nzdgba.co.nz
  www.nzdgba.co.nz

International Arapawa Goat Association
www.arapawagoats.com

Goat-Oriented YahooGroups

Goat Medical Lists

Goat Aid
http://pets.groups.yahoo.com/group/Goat_Aid

Goat ER
http://finance.groups.yahoo.com/group/GoatER

Holistic Goats
http://tech.groups.yahoo.com/group/Holistic-Goats

Home Dairy Goat 911
http://pets.groups.yahoo.com/group/HDG911

Med-A-Goat 911
http://finance.groups.yahoo.com/group/Med-A-Goat911

This and That and Everything Else

Basic Cheesemaking
http://pets.groups.yahoo.com/group/BASICCHEESEMAKING

Goats 101
http://tech.groups.yahoo.com/group/Goats_101

Home Dairy Goats
http://tech.groups.yahoo.com/group/homedairygoats

I'm Kidding
http://pets.groups.yahoo.com/group/Im_Kidding

Not Your Usual Goat List
http://groups.yahoo.com/group/NotYourUsualGoatList
    This is the author's group — please join!

Practical Goats
http://pets.groups.yahoo.com/group/practical-goats

Working Goats

Cart Wagon Goats
http://pets.groups.yahoo.com/group/Cart_Wagon_Goats

Goat Tricks
http://pets.groups.yahoo.com/group/GoatTricks

Packgoat: All Things Packgoat
http://groups.yahoo.com/group/packgoat

Goat Supplies

Camelid Dynamics
    Marty McGee Bennett
    Bend, Oregon
    541-318-5026
www.camelidynamics.com
    Zephyr halters and leads, perfect for goats

Caprine Supply
    De Soto, Kansas
    800-646-7736
www.caprinesupply.com

Gentle Spirit Behavior and Training for Llamas and Alpacas
    Cathy Spalding
cathy@gentlespiritllamas.com
www.gentlespiritllamas.com
    Gentle Spirit Adjustable Halters, perfect for goats

Hamby Dairy Supply
    Maysville, Missouri
    800-306-8937
http://hambydairysupply.com

Hoegger Supply Company
    Fayetteville, Georgia
    800-221-4628
www.hoeggergoatsupply.com

Matilda Sheepcovers
sheepcovers1@bigpond.com

www.sheepcovers.com.au
    Agents located in Australia and the United States

Premier1 Supplies
    Washington, Iowa
    800-282-6631
www.premier1supplies.com

Rocky Sheep Company
    Loveland, Colorado
    970-622-9965
www.rockysheep.com

Sera Inc.
    Central Biomedia Inc.
    Shawnee Mission, Kansas
    913-541-1307
www.seramune.com
    Manufactures Goat Serum Concentrate

Livestock Supplies

Jeffers Livestock
    Dothan, Alabama
    800-533-3377
www.jefferslivestock.com

Port-a-Hut Inc.
    Storm Lake, Iowa
    800-882-4884
www.port-a-hut.com

Quality Llama Products and Alternate Livestock Supply
    Lebanon, Oregon
    800-636-4689
www.llamaproducts.com

Valley Vet Supply
    Marysville, Kansas
    800-419-9524
www.valleyvet.com

References

Aristotle. *The History of Animals*. Translated by D'Arcy Wentworth Thompson. Oxford: Clarendon Press, 1910.

Baldwin, Harold. *Holding the Line*. Chicago: A. C. McClurg & Son, 1919.

Columella. *De Re Rustica*. Translated by Harrison Boyd Ash. Harvard: Loeb Classical Library, 1941.

Dasent, George Webb. *A Selection from the Norse Tales for the Use of Children*. Edinburgh: Edmonston and Douglas, 1862.

Fernie, W. T. *Animal Simples Approved for Modern Uses of Cure*. Bristol: John Wright & Co., 1899

Frazer, Sir James George. *The Golden Bough: A Study in Myth and Religion*. London: MacMillan, 1920.

Oswell, Kate Forrest. *Everychild's Series: The Fairy Book*. New York: MacMillan, 1912.

Pliny. *The Natural History of Pliny*. Translated by John Bostock. London: George Bell and Sons, 1898.

Proceedings of the Royal Irish Academy, vol. 7. Dublin: M. H. Gill, 1862.

Rouse, William Henry Denham. *The Talking Thrush and Other Tales from India*. New York: E. P. Dutton, 1922.

Sikes, Wirt. *British Goblins: Welsh Folk-lore, Fairy Mythology, Legends and Traditions*. London: S. Low, Marston, Searle & Rivington, 1880.

Taylor, Charles R. "Thermoregulatory Functions of the Horns of the Family Bovidae." Cambridge, MA: Harvard University Press, 1966.

Miscellaneous Resources

21st Battalion CEF
http://21stbattalion.ca
 Information on the battalion's mascot, Nan

21sters: 21st Battalion CEF Discussion Group
http://groups.yahoo.com/group/21sters

American Livestock Breeds Conservancy
www.albc-usa.org

The BoerGoats.com Library
www.boergoats.com/clean/library.php
 A good source of goat information and articles, including Nanny Berries. If you love goats, you *must* read Connie Reynolds's sometimes funny, sometimes heartwarming Nanny Berries stories.

Ches McCartney — America's Goat Man
http://the goatman.com
 It's the original Goat Man website!

GoatGenetics.com
 Christine Ball
 Staffordshire, United Kingdom
www.goatgenetics.com
 Specializes in exporting British goat semen and embryos — if you're a North American who fancies British breeds, check it out!

Names by Chinaroad
www.lowchensaustralia.com/Names.htm
 The ultimate guide to great names for goats (and everything else)

The Mopple Chronicles
http://themopplechronicles.blogspot.com
 The Mopple Chronicles is my sheep-training blog, where I also discuss working with goats.

Packing

Equipment

Get Your Goat Gear
St. Anthony, Idaho
877-391-9838
www.getyourgoatgear.com
Goatpacking gear

Hoegger Supply Company
Fayetteville, Georgia
800-221-4628
www.hoeggergoatsupply.com
Goatpacking gear and books

Northwest Pack Goats & Supplies
Weippe, Idaho
888-788-54622
www.northwestpackgoats.com
Training tips, goat packing gear, and books

Quality Llama Products and Alternative Livestock Supply
Lebanon, Oregon
800-638-4689
www.llamaproducts.com

Information

Goat Tracks Magazine
shannon@goattracksmagazine.com
www.goattracksmagazine.com

High Uinta Pack Goats
877-722-5462
www.highuintapackgoats.com
Packgoat trips and extensive information

The Packgoat Forum
www.packgoatforum.com

Summit Pack Goat
Tekamah, Nebraska
402-374-1317
www.summitpackgoat.com
Packgoat trips and extensive information

Organizations

Canadian Pack and Harness Goat Association
packharnessgoats@gmail.com
http://cphga.com

North American Packgoat Association
Boise, Idaho
www.napga.org

Periodicals

Dairy Goat Journal
800-551-5691
www.dairygoatjournal.com

The Goat Magazine
325-653-5438
www.goatmagazine.info

The Goat Rancher
888-562-9529
www.goatrancher.com

Goats Across Canada Magazine
Summer Waters Farm Ltd.
250-569-7406
www.goatsacrosscanada.ca

Ruminations Magazine
978-827-1305
www.smallfarmgoat.com

United Caprine News
817-297-3411
www.unitedcaprinenews.com

Recommended Reading

Beberness, Alice, and Carolyn Eddy. *Field First Aid for Goats*. Eagle Creek, OR. Eagle Creek Packgoats, 2008.

———. *Goat First Aid: The Trail Guide*. Estacade, OR. Eagle Creek Packgoats, 2008.

Belanger, Jerry. *Storey's Guide to Raising Dairy Goats*, 4th ed. North Adams, MA: Storey Publishing, 2010.

Boldrick, Lorrie, and Lydia Hale. *Pygmy Goats: Management and Veterinary Care*. Orange, CA: All Publishing, 1996.

Carroll, Ricki. *Home Cheese Making*, 3rd ed. North Adams, MA: Storey Publishing, 2002.

Damerow, Gail. *Fences for Pasture and Garden*. North Adams, MA: Storey Publishing, 1992.

Damerow, Gail. *Your Goat: A Kid's Guide to Raising and Showing*. North Adams, MA: Storey Publishing, 1993.

Eddy, Carolyn. *Diet for Wethers*. Estacade, OR: Eagle Creek Packgoats, 2001.

———. *Practical Goatpacking*. Estacade, OR. Eagle Creek Packgoats, 1999.

Ekarius, Carol. *How to Build Animal Housing*. North Adams, MA: Storey Publishing, 2004.

———. *Storey's Illustrated Breed Guide to Sheep, Goats, Cattle, and Pigs*. North Adams, MA: Storey Publishing, 2008.

Knight, Anthony P., and Richard G. Walter. *A Guide to Plant Poisoning of Animals in North America*. Jackson, WY: Teton NewMedia, 2001.

Leigh, Jody. *Nigerian Dwarfs: Colorful Miniature Dairy Goats*. Golden, CO: Leighstar Publications, 1993.

Pryor, Karen. *Don't Shoot the Dog! The New Art of Teaching and Training*, rev. ed. Lydney, England: Ringpress Books, 2002.

Sayer, Maggie. *Storey's Guide to Raising Meat Goats*, 2nd ed. North Adams, MA: Storey Publishing, 2010.

Smith, Cheryl K. *Goat Health Care: The Best of Ruminations 2001–2007*. Cheshire, OR: Karmadillo Press, 2009.

Smith, Mary C., and David M. Sherman. *Goat Medicine*, 2nd ed. Ames, IA: Wiley-Blackwell, 2009.

Stewart, Patricia Garland. *Personal Milkers: A Primer to Nigerian Dwarf Goats*. Ashburnham, MA: Garland-Stewart Publishing, 2005.

Tillman, Peggy. *Clicking with Your Dog: Step-by-Step in Pictures*. Waltham, MA: Sunshine Books, 2000.

Waite, Brian. *William de Goat*. Athena Press, 2008.

Weathers, Shirley A. *Field Guide to Plants Poisonous to Livestock — Western U.S.* Fruitland, UT: Rosebud Press, 1998.

Weaver, Sue. *Storey's Guide to Raising Miniature Livestock*. North Adams, MA: Storey Publishing, 2010.

Winslow, Ellie. *The Complete Idiot's Guide to Raising Goats*. New York: Alpha Books, 2010.

———. *Making Money with Goats*, 5th ed. Amarillo, TX: Freefall Press, 2005.

Veterinary

American Association of Small Ruminant Practitioners
334-517-1233
www.aasrp.org

American Holistic Veterinary Medical Association
401-569-0795
www.ahvma.org

American Veterinary Medical Association
800-248-2862
www.avma.org

Glossary

abomasum. The third compartment of the ruminant stomach; the compartment where digestion takes place.

afterbirth. The placenta and any fetal membranes expelled from a doe after kidding.

American Livestock Breeds Conservancy (ALBC). A group dedicated to preserving and promoting rare and endangered breeds of livestock and poultry.

ammonium chloride. A mineral salt fed to male goats to inhibit the formation of bladder and kidney stones.

anestrus. The period of time when a doe is not having the estrus (heat) portion of her estrous cycle.

anthelmintic. A substance used to control or destroy internal parasites; a dewormer.

antibodies. Circulating protein molecules that help neutralize disease organisms.

antitoxin. An antibody capable of neutralizing a specific disease organism.

artificial insemination (AI). A process by which semen is deposited within a doe's uterus by artificial means.

autogenous vaccine. A vaccine made from organisms collected from a specific disease outbreak; e.g., autogenous caseous lymphadenitis vaccine is manufactured using bacteria harvested from pus collected from the lanced abscess of an infected goat.

banding. Castration by the process of applying a fat rubber ring to a buckling's scrotum; see also *elastrator*.

billy (slang). An uncastrated male goat; the preferred term is *buck*.

bit. The part of a driving bridle that goes in a goat's mouth. Goat bits are made of smooth metal with a 3-inch (7.5 cm) mouthpiece and 1-inch (2.5 cm) rings on both sides.

bleating. Goat vocalization; also referred to as *calling*.

bloat. Excessive accumulation of gas in a goat's rumen.

bolus. A large, oval pill; also used to describe a chunk of cud.

booster vaccination. A second or multiple vaccinations given to increase a goat's resistance to a specific disease.

Bo-Se. An injectable prescription selenium supplement.

breast collar. On a goat driving harness, the wide strap that passes around the goat's chest.

breech birth. A birth in which the rump of the kid is presented first.

breeching. The rump straps of a driving harness.

breed. Goats of a color, body shape, and other characteristics similar to those of their ancestors, capable of transmitting these characteristics to their own offspring.

bridle. The head part of a driving harness.

broken-mouth. A goat who has lost some of her permanent incisors, usually at five or more years of age.

browse. Morsels of woody plants such as twigs, shoots, and leaves.

buck. An uncastrated male goat.

buckling. An immature, uncastrated male goat; an uncastrated male kid.

buck rag. A cloth rubbed on the scent glands of a buck and presented to a doe to see if she is in heat.

bunting. The act of a kid poking its dam's udder to stimulate milk letdown.

butting. The act of a goat bashing another goat (or a human) with his horns or forehead.

CAE. See *caprine arthritis encephalitis*.

calling. See *bleating.*

caprine. Having to do with goats.

caprine arthritis encephalitis (CAE). A progressively fatal disease of goats.

caseous lymphadenitis (CL). An incurable disease characterized by abscesses over lymph nodes.

castrate. To remove a male's testes.

cattle panel. A very sturdy, large-gauge, welded-wire fence panel; sold in various lengths and heights.

cc. Cubic centimeter; same as a milliliter (ml).

CD/T. Toxoid vaccine used to protect against enterotoxemia (caused by *Clostridium perfringens* types C and D) and tetanus.

cervix. The section of a doe's uterus that protrudes into the vagina; it dilates during birth to enable kids to pass through.

CL. See *caseous lymphadenitis.*

coccidiostat. A chemical substance mixed with feed, bottle-fed milk, or drinking water to control coccidiosis.

colostrum. First milk a doe gives after birth; high in antibodies, this milk protects newborn kids against disease; sometimes incorrectly called *colostrums.*

concentrate. A high-energy, low-fiber, highly digestible feed such as grain.

condition. Amount of fat and muscle tissue on an animal's body.

cover. To breed (a buck covers a doe).

crossbreed. An animal resulting from the mating of two entirely different breeds.

cud. Undigested food regurgitated by a ruminant to be chewed and swallowed again.

cull. To eliminate an animal from a herd or breeding program.

dam. The female parent.

dehorning. The removal of existing horns.

dental pad. An extension of the gums on the front part of the upper jaw; it is a substitute for top front teeth.

deworm. The use of chemicals or herbs to rid an animal of internal parasites.

dewormer. A substance used to rid an animal of internal parasites; See also *anthelmintic.*

disbud. To destroy the emerging horn buds of a kid by applying a red-hot disbudding iron.

drench. To give liquid medicine by mouth; also a liquid medicine given by mouth.

dystocia. Difficulty in giving birth.

elastrator. A pliers-like tool used to apply heavy, O-shaped rubber bands called elastrator bands to a kid's scrotum for castration.

emaciation. Loss of flesh resulting in extreme leanness.

embryo. An animal in the early stage of development before birth; a fertilized egg.

energy. A nutrient category of feed usually expressed as TDN (total digestible nutrients).

entero. A shortened, common name for enterotoxemia.

estrogen. Female sex hormone produced by the ovaries; estrogen is the hormone responsible for the estrus portion of the estrous cycle.

estrous cycle. The doe's entire reproductive cycle.

estrus. The period when a doe is receptive (will mate with a buck; e.g., she is "in heat") and can become pregnant.

ewe. In countries that use sheep terms to describe goats, a doe or female goat.

fainting goat. A common name for a Myotonic goat.

fecal egg count (FEC). The number of worm eggs in a gram of feces; sometimes written as EPG (eggs per gram).

field shelter. A basic shelter with a roof and three sides.

first freshener. A young doe kidding for the first time.

flehmen. To curl the upper lip up and back in order to increase the ability to discern scent.

forage. Grass and the edible parts of browse plants that can be used to feed livestock.

forb. A broad-leafed herbaceous plant (e.g., curly dock, plantain, or dandelion).

free-choice. Available 24 hours a day, 7 days a week; hay, water, and mineral mixes are generally fed free-choice.

freshen. When a doe kids and begins to produce milk.

gestation. The length of pregnancy.

girth (or belly band). The part of a driving harness that goes under the goat's belly to hold the harness in place.

graft. A procedure in which a kid (or kids) is transferred to and raised by a dam who is not her own.

grain. Seeds of cereal crops such as oats, corn, barley, milo, and wheat.

granny (or auntie) doe. A pregnant doe close to kidding who tries to claim another doe's newborn kid.

gummer. An old goat who has lost all of her teeth.

hay. Grass mowed and cured for use as off-season forage.

heart girth. Circumference of the chest immediately behind the front legs.

heat. See *estrus.*

heritability. The degree to which a trait is passed down to offspring.

hermaphrodite. An animal with both male and female sex organs.

heterosis. The tendency for increased performance or superior qualities in a hybrid offspring over his purebred parents; hybrid vigor.

hybrid vigor. See *heterosis.*

hypothermia. A condition characterized by low body temperature.

immunity. Resistance to a specific disease.

immunoglobulins. A class of proteins found in the blood and other bodily fluids of vertebrates that the immune system uses to neutralize foreign objects such as viruses and bacteria.

in kid. Pregnant.

in milk. Lactating.

intramammary infusion. Mastitis medicine inserted directly into a teat through its orifice.

intramuscular (IM). Into muscle.

intravenous (IV). Into a vein.

Johne's disease. A progressively fatal disease of ruminants, including goats.

jug. An approximately 4- × 5-foot pen where a doe and her kids are put for the first 24 to 72 hours after kidding.

ketones. Substances found in the blood of late-term pregnant goats suffering from pregnancy toxemia.

lactation. The period when a doe is giving milk.

lamb. In countries that use sheep terms to describe goats, a kid or baby goat.

larva. Immature stage of an adult parasite; the term applies to insects, ticks, and worms.

legume. A plant such as alfalfa, clover, and lespedeza that is higher in nitrogen than is a grass hay.

libido. Sex drive; the desire to copulate.

lines. The long reins used with a driving harness.

lymph. A clear, watery, sometimes faintly yellowish fluid derived from body tissues; it contains white blood cells and circulates throughout the lymphatic system.

lymph nodes. Any of the small bodies located along the lymphatic vessels, particularly on the neck behind the ears, farther down the neck, and in the flank area, that filter bacteria and foreign particles from lymph fluid.

magic. A widely used, homemade energy supplement made by combining 1 part corn oil, 1 part molasses, and 2 parts Karo syrup.

mastitis. Inflammation of the udder.

milk letdown. Release of milk by the mammary glands.

ml. Milliliter; see *cc.*

monkey-mouth. See *underbite.*

mothering pen. See *jug.*

myotonia congenital. The inherited neuro-muscular condition that causes major muscles in Myotonic goats temporarily to seize up.

myotonic. The preferred name for "fainting goat"; a goat carrying the gene for myotonia congenital (see above).

nanny (slang). A female goat; the preferred term is *doe*.

nematode. A type of internal parasite; a worm.

nymph. An immature insect or tick that lacks developed sex organs.

off feed. Not eating as much as usual.

off label. The use of a drug for a purpose for which it isn't approved.

omasum. The third part of the ruminant stomach, sandwiched between the reticulum and the abomasum. It is where volatile fatty acids and water are absorbed into the bloodstream.

oocyst. A minute pouch or saclike structure containing a fertilized cell of a parasite.

open doe. One who isn't pregnant.

orifice. The opening in the end of a functional teat.

overshot or parrot-mouth. When the lower jaw is shorter than the upper jaw and the teeth hit in back of the dental pad.

over the counter (OTC). A drug available without a prescription.

ovulation. The release of an egg from the ovary.

ovum. An egg; also called an ova or oocyte.

oxytocin. A naturally occurring hormone important in milk letdown and muscle contraction during the birthing process.

paddock. A small, enclosed area used for grazing.

papers. A animal's registration certificate.

parrot-mouth. The upper dental pad extends beyond the lower teeth.

parturition. The act of giving birth.

pathogen. An agent that causes disease, especially a living microorganism such as a bacterium or a virus.

pedigree. A certificate documenting an animal's line of ancestral descent.

percentage. Partbred; a crossbred goat who is at least 50 percent of a specific breed (e.g., percentage Boer or percentage Kiko).

perennial. A plant that doesn't die at the end of its first growing season but returns and regrows from year to year.

pH. A measure of the activity of hydrogen ions in a solution or substance and therefore its acidity or alkalinity.

pharmaceutical. A substance used in the treatment of disease: a drug, medication, or medicine.

pizzle. The urethral process, a stringy-looking structure at the end of a goat's penis.

placenta. See *afterbirth*.

pneumonia. Infection in the lungs.

polled. A natural absence of horns.

postpartum. After giving birth.

predator. An animal that lives by killing and eating other animals.

prepartum. Before giving birth.

probiotic. A living organism used to influence rumen health by assisting in the fermentation process.

progeny. Offspring.

progesterone. A hormone secreted by the ovaries and produced by the placenta during pregnancy.

proliferate. To vastly multiply in numbers, usually over a short span of time.

prolific. Producing more than the usual number of offspring.

protein. A nutrient category of feed used for growth, milk, and repair of body tissue.

puberty. When a goat becomes sexually mature.

pulpy kidney. Another name for enterotoxemia.

purebred. An animal of a recognized breed who is bred only with other animals of the same breed throughout many generations.

quarantine. To isolate or separate an individual from others of his kind; used particularly to separate sick animals from the rest of the herd.

ram. In countries that use sheep terms to describe goats, an uncastrated male goat.

ration. Total feed given an animal during a 24-hour period.

registered animal. An animal who has a registration certificate and number issued by a breed association.

rehydrate. To replace body fluids lost through dehydration.

reticulum. The second chamber of a ruminant's stomach. Fermentation takes place in the rumen (see below) and the reticulum.

roman-nosed. The convex profile of breeds such as the Boer meat goat and the Nubian dairy goat.

rotational grazing (or browsing). Moving grazing or browsing animals from one paddock to another before plant growth in the first is fully depleted.

roughage. Plant fiber.

roundworm. A parasitic worm with an elongated round body.

rumen. The first compartment of the stomach of a ruminant, in which microbes break down the cellulose in plants.

ruminant. An even-toed mammal that chews its cud and has a multi-compartmented stomach.

rumination. The process whereby a cud or bolus of rumen contents is regurgitated, re-chewed, and re-swallowed; "chewing the cud."

rut. The period during which a buck is interested in breeding females.

scouring. Having diarrhea.

scours. Diarrhea.

scrapie. The goat and sheep version of "mad cow disease"; sometimes incorrectly spelled *scrapies*.

scrotum. The external pouch in which a buck's testicles are suspended.

scrub goat. A mixed breed goat used for brush control.

seasonal breeder. A doe that comes in heat only during part of the year; most dairy goats are seasonal breeders.

selection. Choosing superior animals as parents for future generations.

settle. To get pregnant.

shafts (or fils). Wooden or metal poles that run along both sides of a driving goat, which when connected to his harness enable him to pull the cart.

sheath. The outer skin covering protecting a goat's penis.

shoebox kid. A newborn to 10-day-old market kid.

silent heat. In heat but showing no outward signs. See also *estrus*.

sire. The male parent.

slaughter kid. A kid produced specifically for the meat market.

smooth mouth. A goat who has lost all of his permanent incisors, usually occuring at about 7 years of age.

sow-mouth. See *underbite*.

standing heat. The period during estrus (heat) when a doe allows a buck to breed her.

strip cup. A cup used to catch the first few squirts of milk from each teat.

subcutaneous (SQ). Under the skin.

systemic. Affecting the entire body.

tapeworm. A segmented, ribbonlike, intestinal parasite.

team. In driving, two goats hitched side by side.

teaser. A buck who has had his spermatic cords cut or tied (has had a vasectomy); these males cannot impregnate does but have a sex drive. They are used sometimes to check for and/or help bring a doe into heat.

testosterone. A hormone that promotes the development and maintenance of male sexual characteristics.

total digestible nutrients (TDN). A standard system for expressing the energy value of a feed.

trace mineral. A mineral needed in only minute amounts.

trachea. Windpipe, leading from the larynx in the throat to the lungs; conveys air.

trimester. One-third of a pregnancy.

UC. Urinary calculi; mineral salt crystals ("stones") that form in the urinary tract and sometimes block the urethra of a male goat.

udder. The female mammary system.

ultrasound. A procedure in which sound waves are bounced off tissues and organs; widely used to confirm pregnancy in does.

underbite. Lower jaw is longer than the upper and teeth extend forward past the dental pad on upper jaw; also known as monkey-mouth or sow-mouth.

urethral process. The pizzle; a stringy-looking structure at the end of a goat's penis.

USDA. United States Department of Agriculture.

uterus. The female organ in which fetuses develop; the womb.

vagina. The passageway between the female uterus to the outside of her body; the birth canal.

vascular. Pertaining to or provided with vessels; usually refers to veins and arteries.

wether. A castrated male goat.

withdrawal period. After administering drugs, the amount of time during which an animal must not be sent to market to ensure that no drug residues remain in his meat. (Not all drugs require a withdrawal period; check the label.)

yard. In British terminology, a dry lot where animals (including goats) are kept.

yearling. A goat of either sex who is 1 to 2 years of age, or a goat who has cut his first set of incisors.

zoonose. An animal disease that also infects humans.

Interior Photography Credits

 Index

Page references in *italics* indicate illustrations
or photos; references in **bold** indicate charts.

Other Storey Books You Will Enjoy

The Backyard Cow by Sue Weaver
The essential guide to keeping a productive family cow, including basic instructions for beginners and experience-based insights for seasoned dairy farmers.
240 pages. Paper. ISBN 978-1-60342-997-9.

The Backyard Homestead edited by Carleen Madigan
A complete guide to growing and raising the most local food available anywhere — from one's own backyard.
368 pages. Paper. ISBN 978-1-60342-138-6.

The Backyard Homestead Guide to Raising Farm Animals
edited by Gail Damerow
Expert advice on raising healthy, happy, productive farm animals.
360 pages. Paper. ISBN 978-1-60342-969-6.

Storey's Guide to Raising Series.
Everything you need to know to keep your livestock and your profits healthy. All new editions of *Beef Cattle, Sheep, Pigs, Dairy Goats, Meat Goats, Chickens, Ducks, Turkeys, Poultry, Rabbits, Raising Horses, Training Horses,* and *Llamas.* New additions to the series: *Miniature Livestock* and *Keeping Honey Bees.*
Paper and hardcover. Learn more about each title by visiting *www.storey.com.*

Storey's Illustrated Breed Guide to Sheep, Goats, Cattle, and Pigs
by Carol Ekarius
A comprehensive, colorful, and captivating in-depth guide to North America's common and heritage breeds.
320 pages. Paper. ISBN 978-1-60342-036-5.
Hardcover with jacket. ISBN 978-1-60342-037-2.

These and other books from Storey Publishing are available wherever quality books are sold or by calling 1-800-441-5700. Visit us at *www.storey.com* or sign up for our newsletter at *www.storey.com/signup.*

U

udders and teats, *24,* 24–25, 121
urinary calculi, 148–149

V

vaccinations, 152. *See also* shots,
    administering
vehicle care/maintenance, 96
vet, being your own, 159–164
    first-aid kit, 161
    heart rate, checking, 162
    respiration, assessing, 162
    shots, administering, 162–164, *163, 164*
    temperature, taking goat's, 160
    vital signs, checking, 160
veterinarians, 158–159, 178
vital signs, checking, 7, 160

W

wagons. *See* carts and wagons
water
    crossing, 110

goats hating, 15–16
    supply of, 150, 152
watering devices, 143
wattles, 10, *10*
weaning, 14
weighing goats, 152, **153**
wellness. *See* healthy goats
Welsh regimental goats, 13, *13*
wethers, 29, 150, 167
wild goats, 5, *5, 8*
William de Goat, 134
wines and beers, 75
Wood, Lindsay A., 62
World War I, 100–101
World War II, 59, 134
worms. *See* internal parasites

Z

Zeder, Melinda, 4
zoning laws, 18